FINITE DIFFERENCES

A PROBLEM SOLVING TECHNIQUE

**DALE SEYMOUR
PALO ALTO, CALIFORNIA**

**MARGARET SHEDD
BERKELEY HIGH SCHOOL
BERKELEY, CALIFORNIA**

DALE SEYMOUR PUBLICATIONS

When you complete this book see if you can decode the cover.

ISBN 0-86651-350-7
(previously ISBN 0-88488-047-8)
Order number DS01719

DALE
SEYMOUR
PUBLICATIONS
P.O. BOX 10888
PALO ALTO, CA 94303

abcdefghi-MA-893210987

TABLE OF CONTENTS

INTRODUCTION

This book discusses how to solve certain types of mathematical problems using the finite differences technique. Since the technique is unknown to many students and teachers of mathematics, this book has been written to provide a useful introduction. The method is simple to learn, understand and use. It sometimes makes the solution of otherwise difficult problems relatively easy.

The technique of finite differences is explained on pages 31 – 40. It is recommended that you try to solve some of the problems in the first section of the book before you read about the technique.

While solving some of the problems, you may uncover the technique yourself. Surely numerous people have discovered it on their own during the past centuries.

Wherever appropriate, the page format is designed to be used as a worksheet or transparency for classroom use. See the copyright page for limited reproduction rights.

Problems are presented in three sections. In general, they become progressively complex. All but one or two of the problems in the book can be solved using finite differences. In some cases there may be other problem solving techniques which produce solutions more quickly. A good problem-solver learns many approaches. Through practice he learns to apply the most expedient technique for solving a problem.

This book is designed to appeal to students and teachers with a wide range of problem solving skills and math backgrounds. Every reader will not need to work every page. Look through the book to locate sections you can skim and those you need to concentrate on. Select appropriate problems to work.

A glossary of terms used throughout the book is provided on pages 108–110. Words that appear in italics are defined in the glossary.

PROBLEM SECTION ▌

■ **Sequences and Series**

■ **What's My Rule?**

■ **Square Numbers**

■ **Triangular Numbers**

■ **Oblong Numbers**

■ **Pentagonal Numbers**

■ **Hexagonal Numbers**

■ **Diagonals of a Polygon**

■ **Cubic Numbers**

■ **Points and Lines**

■ **Segments on a Line**

■ **Problem Summary**

"There can be no mystery in a result you have discovered for yourself."

W. W. Sawyer

SEQUENCES AND SERIES

A *sequence* is an ordered set of numbers formed according to some pattern.

> Examples: 1, 2, 3, 4, 5, 6, . . .
>
> 5, 10, 15, 20, 25, . . .
>
> 4, 9, 15, 22, 30, . . .

A *series* is an indicated sum of the terms of a sequence.

> Examples: $1 + 2 + 3 + 4 + 5 + 6 + . . .$
>
> $5 + 10 + 15 + 20 + 25 + . . .$
>
> $4 + 9 + 15 + 22 + 30 + . . .$

Each element of the sequence or series is called a term.

> Example: In the sequence 5, 10, 15, 20, . . .
>
> 5 is the first term, 10 is the second term,
>
> 15 is the third term, etc.

The *general term* is an arbitrary term. In this book we shall refer to the general term as the nth *term*. The nth term describes a rule.

> Example: If the nth term of a sequence is $2n + 5$,
>
> the first term (n=1) is $2(1) + 5$ or 7,
>
> the second term (n=2) is $2(2) + 5$ or 9,
>
> the third term (n=3) is $2(3) + 5$ or 11,
>
> the hundredth term (n=100) is $2(100) + 5$ or 205,
>
> the nth term (n=n) is $2n + 5$

In many of the worksheets in this book we shall chart a table of values such as the one shown at the right. In this table n represents the number of the term in sequence (first, second, third, fourth, . . ., nth). The letter T represents the value of the term. The table at the right represents the sequence 7, 9, 11, 13, 15, . . ., $2n + 5$.

n	T
1	7
2	9
3	11
4	13
5	15
.	.
.	.
.	.
n	$2n+5$

WHAT'S MY RULE?

When solving problems, it is often helpful to know a general rule. Finding the nth term in a sequence is finding the general rule for all the terms in that sequence.

Complete each sequence and find the nth term. nth term

1 2, 4, 6, 8, 10, ___, ___, ___, . . .   ___

2 1, 3, 5, 7, 9, ___, ___, ___, . . .   ___

3 3, 6, 9, 12, 15, ___, ___, ___, . . .   ___

4 7, 10, 13, 16, 19, ___, ___, ___, . . .   ___

5 5, 10, 15, 20, 25, ___, ___, ___, . . .   ___

6 2, 7, 12, 17, 22, ___, ___, ___, . . .   ___

7 $\frac{3}{2}$, 2, $\frac{5}{2}$, 3, $\frac{7}{2}$, ___, ___, ___, . . .   ___

8 $-\frac{3}{2}$, -1, $-\frac{1}{2}$, 0, $\frac{1}{2}$, ___, ___, ___, . . .   ___

9 17, 25, 33, 41, 49, ___, ___, ___, . . .   ___

10 9, 23, 37, 51, 65, ___, ___, ___, . . .   ___

9

SQUARE NUMBERS

Square numbers are numbers which can be represented by dots in a square arrangement or array.

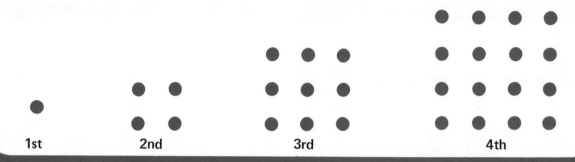

In the table to the right, *n* represents the number of dots on one side of an array. *S* represents the total number of dots in the array.

Complete the table at the right through the eighth square number.

Find a pattern which will enable you to find the 20th and 100th square numbers without drawing or counting them.

How did you find the pattern?

Express the rule for the *n*th square number.

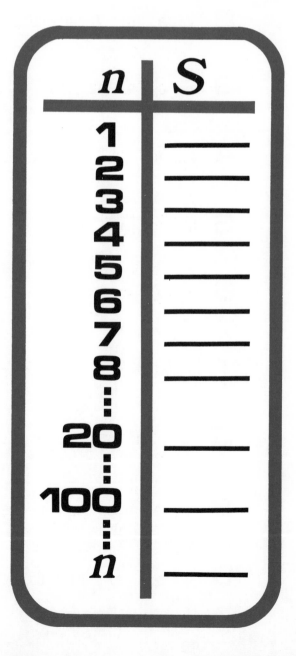

TRIANGULAR NUMBERS

Triangular numbers are numbers which can be represented by dots in an equilateral triangular array.

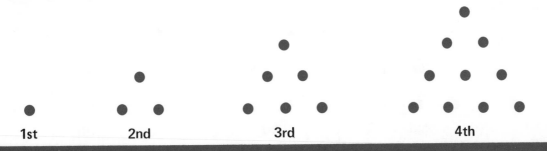

1st   2nd   3rd   4th

In the table to the right, n represents the number of dots on one side of an array. T represents the total number of dots in the array.

Complete the table at the right through the ninth triangular number.

Find a pattern which will enable you to find the 30th and 100th triangular numbers without drawing or counting them.

How did you find the pattern?

Express the rule for the nth triangular number.

n	T
1	
2	
3	
4	
5	
6	
7	
8	
9	
30	
100	
n	

11

OBLONG NUMBERS

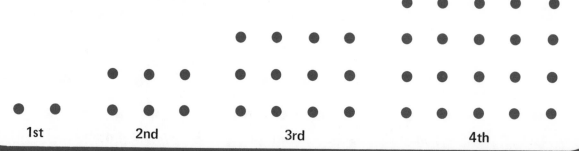

Oblong numbers are numbers which can be represented in a rectangular array having one dimension one unit longer than the other.

1st   2nd   3rd   4th

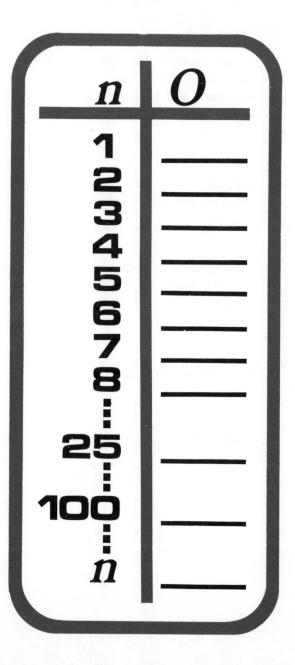

In the table to the right, n represents the smallest dimension of an oblong array. O represents the total number of dots in the array.

Complete the table at the right through the eighth oblong number.

Find a pattern which will enable you to find the 25th and 100th oblong numbers without drawing or counting them.

How did you find the pattern?

Express the nth oblong number as a general rule.

n	O
1	
2	
3	
4	
5	
6	
7	
8	
25	
100	
n	

PENTAGONAL NUMBERS

Pentagonal numbers are numbers which can be represented by dots in a pentagonal array.

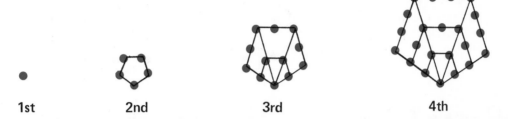

| 1st | 2nd | 3rd | 4th |

In the table to the right, *n* represents the number of dots on one side of the array. *P* represents the total number of dots in the array.

Complete the table at the right through the eighth pentagonal number.

Find a pattern which will enable you to find the tenth and 100th pentagonal numbers without drawing or counting them.

How did you find the pattern?

Express the *n*th pentagonal number as a general rule.

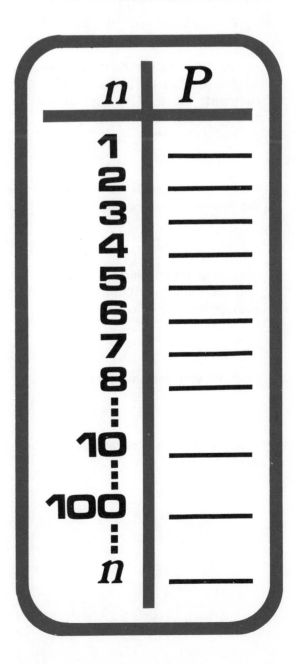

HEXAGONAL NUMBERS

Hexagonal numbers are numbers which can be represented by dots in a hexagonal array.

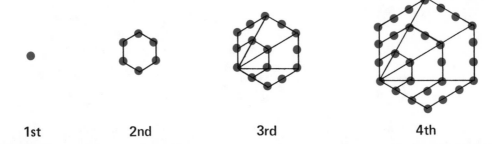

1st   2nd   3rd   4th

In the table to the right, *n* represents the number of dots on one side of an array. *H* represents the total number of dots in the array.

Complete the table at the right through the seventh hexagonal number.

Find a pattern which will enable you to find the 15th and 100th hexagonal numbers without drawing or counting them.

How did you find the pattern?

Express the *n*th hexagonal number as a general rule.

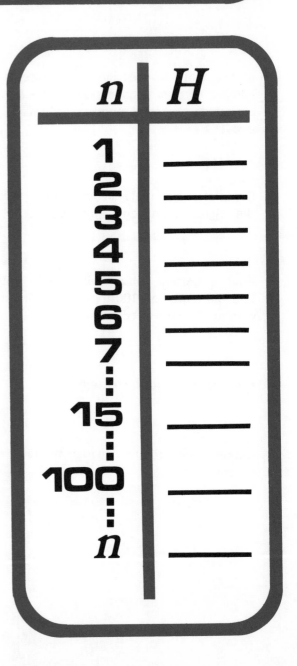

n	H
1	___
2	___
3	___
4	___
5	___
6	___
7	___
15	___
100	___
n	___

DIAGONALS OF A POLYGON

A diagonal of a polygon is a line segment joining any two non-adjacent vertices. Here, n represents the number of sides in the polygon.

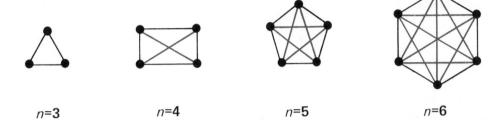

$n=3$ $n=4$ $n=5$ $n=6$

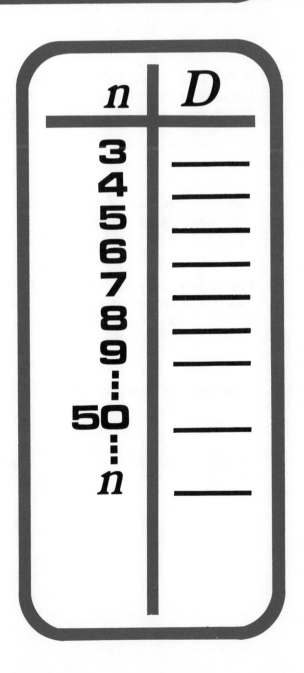

n	D
3	___
4	___
5	___
6	___
7	___
8	___
9	___
...	
50	___
...	
n	___

In the table to the right, D represents the number of diagonals in a given polygon.

Complete the table at the right through a nine-sided polygon.

Find a pattern which will enable you to find the diagonals in a 50-sided polygon without drawing it.

How did you find the pattern?

Express the general rule for finding the number of diagonals in an n-sided polygon.

CUBIC NUMBERS

Cubic numbers are numbers which can be represented by objects in a cubic array.

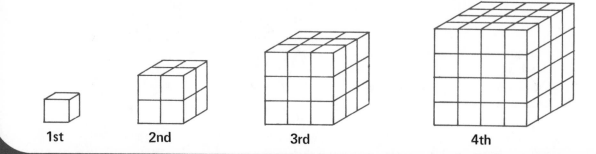

1st   2nd   3rd   4th

In the table to the right, n represents the number of unit cubes on one edge of a cube. C represents the total number of units in a given cube.

Complete the table at the right through the seventh cubic number.

Find a pattern which will enable you to find the 12th and 100th cubic numbers without drawing or counting them.

How did you find the pattern?

Express the nth cubic number as a general rule.

n	C
1	
2	
3	
4	
5	
6	
7	
12	
100	
n	

16

POINTS AND LINES

Non-*collinear* points on a plane are points not in a straight line. They can be connected by *line segments.*   Here, *n* represents the number of points.

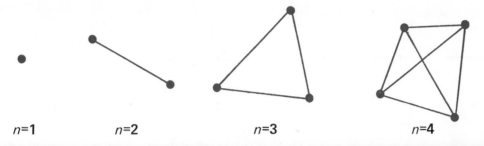

n=1   n=2   n=3   n=4

In the table to the right, *L* represents the number of line segments connecting a given number of points.

Complete the table at the right through ten points.

Find a pattern which will enable you to find the number of line segments that can connect 20 non-collinear points on a plane.

How did you find the pattern?

Express the general rule for the number of line segments connecting *n* non-collinear points on a plane.

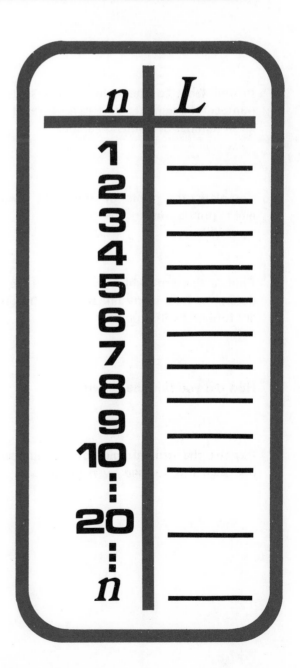

n	L
1	
2	
3	
4	
5	
6	
7	
8	
9	
10	
20	
n	

SEGMENTS ON A LINE

Points on a line establish the line segments on that line. Here, *n* is the number of points on a line.

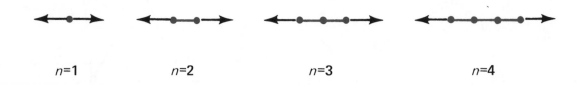

n=1   *n*=2   *n*=3   *n*=4

In the table to the right, *S* represents the number of segments (not rays) formed by a given number of points on a line.

Complete the table at the right through eight points on a line.

Find a pattern which will enable you to determine how many line segments (not rays) are formed by 100 points.

How did you find the pattern?

Express the general rule for the number of line segments formed by *n* points.

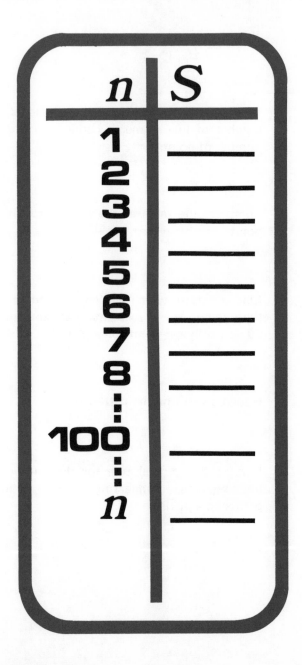

n	*S*
1	
2	
3	
4	
5	
6	
7	
8	
100	
n	

PROBLEM SUMMARY

For each of the following sequences, write the first six terms, the 12th term, and the general, or nth, term.

		12th	nth

1 Square numbers: ___, ___, ___, ___, ___, ___, . . . ,   ___   ___

2 Triangular numbers: ___, ___, ___, ___, ___, ___, . . . ,   ___   ___

3 Oblong numbers: ___, ___, ___, ___, ___, ___, . . . ,   ___   ___

4 Pentagonal numbers: ___, ___, ___, ___, ___, ___, . . . ,   ___   ___

5 Hexagonal numbers: ___, ___, ___, ___, ___, ___, . . . ,   ___   ___

6 Cubic numbers: ___, ___, ___, ___, ___, ___, . . . ,   ___   ___

7 How many diagonals are there in an n-sided polygon? _____

8 How many different line segments (not rays) are formed by n points on a line? _____

9 How many different line segments can be drawn between n non-collinear points? _____

PATTERNS AND TECHNIQUES FOR FINDING THE SOLUTIONS

"In mathematics there is always a pattern."

Peter, age 10

■ **Carl Friedrich Gauss**

■ **Searching for Patterns**

CARL FRIEDRICH GAUSS

If you have solved all or most of the problems in Section I of this book, you are an excellent problem solver. Many of the problems require considerable insight into number patterns to discover the correct solutions.

You may have used problem solving skills that you learned earlier to solve some of the problems. Or, if these kinds of problems were new to you, you may have found some patterns that helped you find solutions.

Carl Friedrich Gauss, one of the greatest mathematicians of all time, was discovering mathematical patterns when he was 11. The anecdote goes that in Germany in about 1787 a headmaster gave his class of 11 year olds a problem to keep them busy while he worked. He told the class to find the sum of all the numbers from 1-100. Although annoyed by this tedious task, the students were too afraid of their strict teacher to object. They all began the laboring addition - except one. Carl Friedrich Gauss thought there must be an easier way to compute the answer than to write down all of the numbers. He looked for a pattern. Imagine the headmaster's surprise when Gauss presented him with the correct answer moments after the problem was posed.

Can you figure out a method to compute the answer?

Gauss used one of the two methods shown below, but we are not sure which one.

METHOD 1

List the numbers in order:

1 + 2 + 3 + 4 + 5 + . . . + 97 + 98 + 99 + 100

Pair the first and last, the second and next to last, etc. Find the sum of each pair.

1 + 2 + 3 + 4 + 5 + . . . + 97 + 98 + 99 + 100

101

101

101

101

This results in 50 pairs, each with a sum of 101. The sum of the numbers from 1 to 100 is then, 50 x 101 or 5050.

METHOD 2

List the series from 1 to 100. Below that, list the series from 100 to 1. Add each pair of numbers.

$$1 + 2 + 3 + 4 + 5 + \ldots + 97 + 98 + 99 + 100$$
$$\underline{100} + \underline{99} + \underline{98} + \underline{97} + \underline{96} + \ldots + \underline{4} + \underline{3} + \underline{2} + \underline{1}$$
$$101 \quad 101 \quad 101 \quad 101 \quad 101 \qquad 101 \quad 101 \quad 101 \quad 101$$

This results in 100 sums of 101. Half of these pairs contain all the numbers from 1 to 100. The sum of all the numbers from 1 to 100 is then, $\frac{100}{2}$ x 101 or 5050.

These are ways to find the sum of a special series such as the *counting numbers*. In general, the sum of n counting numbers is ½$n(n+1)$. Because the pattern of counting numbers occurs frequently in other problems, it is helpful to remember this technique. The formula will be useful.

More often than not, there is more than one way to solve a problem. This section will briefly point out some ways which might be used to "unlock" the mysteries of problems in Section I.

You will appreciate these patterns for solving the problems much more if you have already tried to solve the problems yourself. Possibly, after reading about some of the pattern finding techniques, you will want to go back to certain problems in Section I and see if these techniques work.

One technique that can be used is finding common differences. This method will be introduced in this section and described in detail in the following section. You may want to solve some of the problems in Section I using your own patterns before the common differences technique is revealed in detail. Maybe, you will experience the excitement of deriving the method yourself.

SEARCHING FOR PATTERNS

METHOD 1 RECOGNIZING SPECIAL NUMBERS

Sometimes in a sequence of numbers you will recognize that all of the numbers share some special property. Some examples are given below.

Sequence	Property	Rule
1, 4, 9, 16, 25, 36, 49,	All numbers are square numbers	n^2
1, 8, 27, 64, 125,	All numbers are cube numbers	n^3
3, 6, 9, 12, 15, 18	All numbers are multiples of three	$3n$
1, 3, 5, 7, 9, 11	All numbers are consecutive odd numbers	$(2n-1)$

As problems become more complex it is not likely that you will be able to recognize the relationship immediately by simply observing the numbers. You will probably need to look for additional types of patterns.

METHOD 2 THE TERM AS A SERIES

It may be helpful to break the numbers of the sequence into a series of numbers which reveal a pattern. For example, let's look at the problem of finding the nth pentagonal number.

Pentagonal numbers

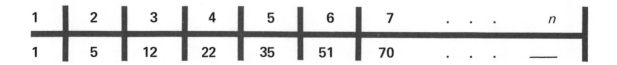

1	2	3	4	5	6	7	. . .	n
1	5	12	22	35	51	70	. . .	___

The table could be written as follows:

n	P	
1	1	(1)
2	5	(1+4)
3	12	(1+4+7)
4	22	(1+4+7+10)
5	35	(1+4+7+10+13)
6	51	(1+4+7+10+13+16)
⋮	⋮	
n		(1+4+7+10+13+16+. . .+ ___)

Noticing a difference of 3 between terms, the numbers could be written in the following way.

n	P	
1	1	(1)
2	5	(1+1+3)
3	12	(1+1+3+1+2·3)
4	22	(1+1+3+1+2·3+1+3·3)
5	35	(1+1+3+1+2·3+1+3·3+1+4·3)
6	51	(1+1+3+1+2·3+1+3·3+1+4·3+1+5·3)
⋮	⋮	
n		[1+1+3+1+2·3+1+3·3+. . .+1+(n-1)3]

The pattern could also be written in this way:

n	P	
1	1	(1·1)
2	5	(2·1) + (1·3)
3	12	(3·1) + (1·3) + (2·3)
4	22	(4·1) + (1·3) + (2·3) + (3·3)
5	35	(5·1) + (1·3) + (2·3) + (3·3) + (4·3)
6	51	(6·1) + (1·3) + (2·3) + (3·3) + (4·3) + (5·3)
⋮	⋮	
n		(n·1) + (1·3) + (2·3) + (3·3) + . . . + (n-1)3

Using the distributive property the numbers could be written like this:

n	P	
1	1	$1 + 3(0)$
2	5	$2 + 3(1)$
3	12	$3 + 3(1+2)$
4	22	$4 + 3(1+2+3)$
5	35	$5 + 3(1+2+3+4)$
6	51	$6 + 3(1+2+3+4+5)$
.	.	
.	.	
.	.	
n		$n + 3[(1+2+3+4+\ldots+(n-1)]$ or $n+3[\frac{n(n-1)}{2}]$ or $\frac{n(3n-1)}{2}$

Once again the series of counting numbers appears.

The general rule was found by writing each term of the sequence as a series of numbers and discovering a pattern in the way the terms progressed.

METHOD 3 FINDING A PREVIOUSLY KNOWN RULE

The pattern in the last example contained a previously known sequence, the triangular numbers. Since a rule is already known for the triangular numbers, it can be used to develop a generalization.

The pentagonal numbers can be written as follows:

n	P	
1	1	$1 + 3(0)$
2	5	$2 + 3(1)$
3	12	$3 + 3(3)$
4	22	$4 + 3(6)$
5	35	$5 + 3(10)$
6	51	$6 + 3(15)$
7	70	$7 + 3(21)$
8	92	$8 + 3(28)$
.	.	
.	.	
.	.	
n		$n+3(n\text{-}1)$th triangular number

The pattern is n plus three times the $(n-1)$th triangular number. Since the nth triangular number is $\frac{n(n+1)}{2}$, then the $(n-1)$th triangular number is $\frac{(n-1)(n-1+1)}{2}$ or $\frac{(n-1)n}{2}$. So the rule for pentagonal numbers is $n+3[\frac{(n-1)n}{2}]$ or $\frac{n(3n-1)}{2}$.

METHOD 4 OBSERVING A VISUAL PATTERN

Sometimes a geometric pattern may become obvious, as in the pentagonal numbers shown below.

n	P		
1	1	1st square no. + 0th triangular no.	•
2	5	2nd square no. + 1st triangular no.	
3	12	3rd square no. + 2nd triangular no.	
4	22	4th square no. + 3rd triangular no.	
5	35	5th square no. + 4th triangular no.	
6	51	6th square no. + 5th triangular no.	
⋮	⋮		
n		nth square no. + (n-1)th triangular no. or $n^2 + \frac{(n-1)n}{2}$ equals $\frac{n(3n-1)}{2}$	

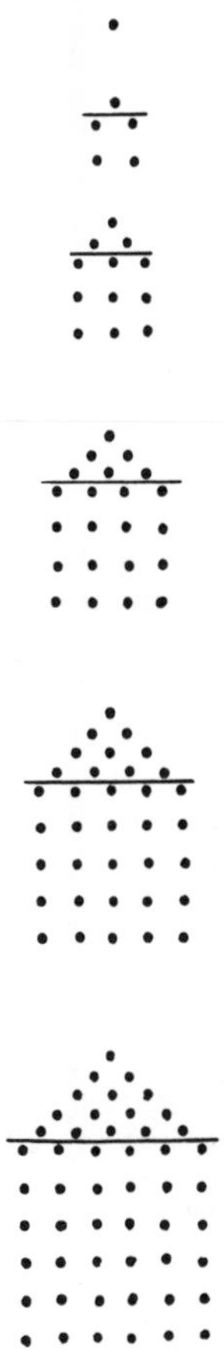

This is the same rule we discovered using methods 1 and 2.

METHOD 5 FINITE DIFFERENCES

In many of the problems in the first part of Section I successive terms differed by a constant number. For example:

$$2, \quad 7, \quad 12, \quad 17, \quad 22, \quad \text{\_\_\_}, \quad \text{\_\_\_}, \quad \text{\_\_\_}, \ldots$$

$$\lor \quad \lor \quad \lor \quad \lor \quad \lor \quad \lor \quad \lor \quad \lor$$
$$5 \quad 5 \quad 5 \quad 5 \quad 5 \quad 5 \quad 5 \quad 5$$

This does not happen when you subtract the successive terms of the pentagonal number sequence.

n	P
1	1
	$>$ 4
2	5
	$>$ 7
3	12
	$>$ 10
4	22
	$>$ 13
5	35
	$>$ 16
6	51

The differences do have a pattern, however. They differ by 3. If we subtract the differences this becomes obvious.

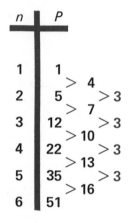

n	P
1	1
	$>$ 4
2	5	$>$ 3
	$>$ 7
3	12	$>$ 3
	$>$ 10
4	22	$>$ 3
	$>$ 13
5	35	$>$ 3
	$>$ 16
6	51

Is this just a coincidence or could this pattern help us find a rule? Could there be a similar pattern in the problems from Section I? Let's test some examples.

Square Numbers

n	S
1	1
	$>$ 3
2	4	$>$ 2
	$>$ 5
3	9	$>$ 2
	$>$ 7
4	16	$>$ 2
	$>$ 9
5	25	$>$ 2
	$>$ 11
6	36

Triangular Numbers

n	T
1	1
	$>$ 2
2	3	$>$ 1
	$>$ 3
3	6	$>$ 1
	$>$ 4
4	10	$>$ 1
	$>$ 5
5	15	$>$ 1
	$>$ 6
6	21

Oblong Numbers

n	O		
1	2		
		> 4	
2	6		> 2
		> 6	
3	12		> 2
		> 8	
4	20		> 2
		> 10	
5	30		> 2
		> 12	
6	42		

Hexagonal Numbers

n	H		
1	1		
		> 5	
2	6		> 4
		> 9	
3	15		> 4
		> 13	
4	28		> 4
		> 17	
5	45		> 4
		> 21	
6	66		

It does look like the common difference is more than a coincidence. Notice that in the problem about the number of squares in a square, it was necessary to subtract three times to get a common difference. These patterns can lead to a solution of the problem. One advantage of this approach is that you don't have to wait for an inspirational insight. Simply subtract terms of differences until a common difference appears. This is the technique called *finite differences.* The next section will explain how it can be used for any term in a patterned sequence of numbers. But remember that this method doesn't lead to the solution of all sequence problems, nor is it always the best way to approach a sequence problem. It is, however, a very handy problem solving tool with which you should be familiar.

At this point, you may wish to see if you can discover the method yourself without reading the next section. This would be a very gratifying accomplishment. Certainly thousands before you have discovered the method on their own.

THE TECHNIQUE OF FINITE DIFFERENCES

"What is the difference between method and device? A method is a device which you can use twice."

G. Polya

- **Common Difference Pattern**

- **Generating a Sequence**

- **Summary**

- **Some Limitations**

COMMON DIFFERENCE PATTERN

In the last section on patterns, we saw that by subtracting successive terms in a sequence we found a new sequence. If we keep subtracting, we may eventually get a common difference. This table of triangular numbers is an example.

n	T	1st diff.	2nd diff.
1	1		
		> 2	
2	3		> 1
		> 3	
3	6		> 1
		> 4	
4	10		> 1
		> 5	
5	15		> 1
		> 6	
6	21		
.	.		
.	.		
.	.		
n	$\dfrac{n(n+1)}{2}$		

This kind of pattern occurred in nearly every problem in Section I. We have reason to suspect that if we knew more about why these patterns occur, we might discover a technique for finding rules about number sequences. Our approach has been to take a sequence and look for a pattern so that we may discover the general rule.

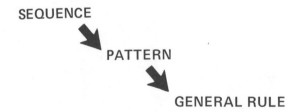

SEQUENCE

PATTERN

GENERAL RULE

Sometimes it is helpful, when solving problems, to work backwards. That is, start with a rule and find the sequence. Either way, finding patterns will help disclose a problem solving technique.

GENERATING A SEQUENCE

We can chart a table of values from a rule by substituting the natural numbers (1, 2, 3, 4, 5, . . .) for n in the rule. Given some rules like $n+3$, $2n-1$, $4n+5$, $7n-8$ the tables would be:

n	T
1	4
2	5
3	6
4	7
5	8
6	9
.	.
.	.
.	.
n	$n+3$

n	T
1	1
2	3
3	5
4	7
5	9
6	11
.	.
.	.
.	.
n	$2n-1$

n	T
1	9
2	13
3	17
4	21
5	25
6	29
.	.
.	.
.	.
n	$4n+5$

n	T
1	−1
2	6
3	13
4	20
5	27
6	34
.	.
.	.
.	.
n	$7n-8$

If we subtract successive terms in these sequences, the tables are:

4	
	> 1
5	
	> 1
6	
	> 1
7	
	> 1
8	
	> 1
9	
.	
.	
.	
$1n+3$	

1	
	> 2
3	
	> 2
5	
	> 2
7	
	> 2
9	
	> 2
11	
.	
.	
.	
$2n-1$	

9	
	> 4
13	
	> 4
17	
	> 4
21	
	> 4
25	
	> 4
29	
.	
.	
.	
$4n+5$	

−1	
	> 7
6	
	> 7
13	
	> 7
20	
	> 7
27	
	> 7
34	
.	
.	
.	
$7n-8$	

Let's look at some more rules and their tables.

n^2, $n^2 + 2n + 3$, $n^2 - n + 1$, $n^2 + 7$

n	T
1	1
2	4
3	9
4	16
5	25
6	36
.	.
.	.
.	.
n	n^2

n	T
1	6
2	11
3	18
4	27
5	38
6	51
.	.
.	.
.	.
n	$n^2 + 2n + 3$

n	T
1	1
2	3
3	7
4	13
5	21
6	31
.	.
.	.
.	.
n	$n^2 - n + 1$

n	T
1	8
2	11
3	16
4	23
5	32
6	43
.	.
.	.
.	.
n	$n^2 + 7$

Subtracting successive terms, the tables are:

```
1                    6                    1                    8
   > 3                  > 5                  > 2                  > 3
4        > 2         11        > 2         3        > 2        11        > 2
   > 5                  > 7                  > 4                  > 5
9        > 2         18        > 2         7        > 2        16        > 2
   > 7                  > 9                  > 6                  > 7
16        > 2         27        > 2         13        > 2        23        > 2
   > 9                  > 11                 > 8                  > 9
25        > 2         38        > 2         21        > 2        32        > 2
   > 11                 > 13                 > 10                 > 11
36                   51                   31                   43

  .                    .                    .                    .
  .                    .                    .                    .
  .                    .                    .                    .

$n^2$              $n^2 + 2n + 3$        $n^2 - n + 1$        $n^2 + 7$
```

Now let's try some rules, and their tables, where the *numerical coefficient* of the n^2 term is 2.

$2n^2$; $2n^2 - n + 1$; $2n^2 - 3n$; $2n^2 + 5$

n	T
1	2
2	8
3	18
4	32
5	50
6	72
⋮	⋮
n	$2n^2$

n	T
1	2
2	7
3	16
4	29
5	46
6	67
⋮	⋮
n	$2n^2 - n + 1$

n	T
1	−1
2	2
3	9
4	20
5	35
6	54
⋮	⋮
n	$2n^2 - 3n$

n	T
1	7
2	13
3	23
4	37
5	55
6	77
⋮	⋮
n	$2n^2 + 5$

Subtracting successive terms, the tables are:

```
2                    2                    −1                   7
   > 6                  > 5                  > 3                  > 6
8        > 4         7        > 4         2        > 4         13        > 4
   > 10                 > 9                  > 7                  > 10
18        > 4         16        > 4         9        > 4         23        > 4
   > 14                 > 13                 > 11                 > 14
32        > 4         29        > 4         20        > 4         37        > 4
   > 18                 > 17                 > 15                 > 18
50        > 4         46        > 4         35        > 4         55        > 4
   > 22                 > 21                 > 19                 > 22
72                   67                   54                   77

⋮                    ⋮                    ⋮                    ⋮

$2n^2$              $2n^2 - n + 1$       $2n^2 - 3n$          $2n^2 + 5$
```

The common difference seems to be <u>twice</u> the coefficient of the n^2 term.

Let's try one more set of varied examples to see if the common difference is twice the coefficient of the n^2 term in these also. If the pattern holds, the common difference will be

6 for $3n^2 + 1$

10 for $5n^2 - n + 1$

8 for $4n^2 - 3$

14 for $7n^2 + 2n - 1$

The tables show:

n	T		
1	4		
		> 9	
2	13		> 6
		> 15	
3	28		> 6
		> 21	
4	49		> 6
		> 27	
5	76		> 6
		> 33	
6	109		
⋮	⋮		
n	$3n^2 + 1$		

n	T		
1	5		
		> 14	
2	19		> 10
		> 24	
3	43		> 10
		> 34	
4	77		> 10
		> 44	
5	121		> 10
		> 54	
6	175		
⋮	⋮		
n	$5n^2 - n + 1$		

n	T		
1	1		
		> 12	
2	13		> 8
		> 20	
3	33		> 8
		> 28	
4	61		> 8
		> 36	
5	97		> 8
		> 44	
6	141		
⋮	⋮		
n	$4n^2 - 3$		

n	T		
1	8		
		> 23	
2	31		> 14
		> 37	
3	68		> 14
		> 51	
4	119		> 14
		> 65	
5	184		> 14
		> 79	
6	263		
⋮	⋮		
n	$7n^2 + 2n - 1$		

We seem to have found a pattern. The common difference is twice the coefficient of the n^2 term. But how do we know it will <u>always</u> follow this pattern? We can test the general formula for a quadratic equation.

A *polynomial* is said to be quadratic (second degree) if it can be written in the form $an^2 + bn + c$ where a, b, and c are any real numbers. The expressions listed below, which we have been testing, are all examples of quadratic equations.

expression	$an^2 + bn + c$	a, b, c
n^2	$1n^2 + 0n + 0$	$a = 1, b = 0, \ c = 0$
$n^2 + 2n + 3$	$1n^2 + 2n + 3$	$a = 1, b = 2, \ c = 3$
$n^2 - n + 1$	$1n^2 - 1n + 1$	$a = 1, b = -1, c = 1$
$n^2 + 7$	$1n^2 + 0n + 7$	$a = 1, b = 0, \ c = 7$
$2n^2$	$2n^2 + 0n + 0$	$a = 2, b = 0, \ c = 0$
$2n^2 - n + 1$	$2n^2 - 1n + 1$	$a = 2, b = -1, c = 1$
$2n^2 - 3n$	$2n^2 - 3n + 0$	$a = 2, b = -3, c = 0$
$2n^2 + 5$	$2n^2 + 0n + 5$	$a = 2, b = 0, \ c = 5$
$3n^2 + 1$	$3n^2 + 0n + 1$	$a = 3, b = 0, \ c = 1$

Let's substitute some values in a table for the general quadratic expression $an^2 + bn + c$. This should show what patterns to expect on any second degree expression.

n	$an^2 + bn + c$
0	c
1	$a + b + c$
2	$4a + 2b + c$
3	$9a + 3b + c$
4	$16a + 4b + c$
5	$25a + 5b + c$
6	$36a + 6b + c$
.	
.	
.	
n	$an^2 + bn + c$

Now, if we subtract to find the common differences, we get

n	$an^2 + bn + c$		
0	c		
		$> a + b$	
1	$a + b + c$		$> 2a$
		$> 3a + b$	
2	$4a + 2b + c$		$> 2a$
		$> 5a + b$	
3	$9a + 3b + c$		$> 2a$
		$> 7a + b$	
4	$16a + 4b + c$		$> 2a$
		$> 9a + b$	
5	$25a + 5b + c$		$> 2a$
		$> 11a + b$	
6	$36a + 6b + c$		
.			
.			
.			
n	$an^2 + bn + c$		

We get a common difference in the second subtraction: second subtraction—second degree expression. That makes it easy to remember. We also see that we were correct about the observed pattern. The common difference ($2a$) is always going to be twice the coefficient of the n^2 term (a).

Now let's see how we could use this table of general values to find the rule for a sequence of numbers which has a common difference in the second series of subtractions.

Let's take the example $2n^2 + 3n - 4$. Compare the tables below.

n	$2n^2 + 3n - 4$
0	-4
	> 5
1	1 > 4
	> 9
2	10 > 4
	> 13
3	23 > 4
	> 17
4	40 > 4
	> 21
5	61 > 4
	> 25
6	86
.	.
.	.
.	.
n	$2n^2 + 3n - 4$

n	$an^2 + bn + c$
0	c
	$> a + b$
1	$a + b + c$ $> 2a$
	$> 3a + b$
2	$4a + 2b + c$ $> 2a$
	$> 5a + b$
3	$9a + 3b + c$ $> 2a$
	$> 7a + b$
4	$16a + 4b + c$ $> 2a$
	$> 9a + b$
5	$25a + 5b + c$ $> 2a$
	$> 11a + b$
6	$36a + 6b + c$
.	.
.	.
.	.
n	$an^2 + bn + c$

Compare these values:

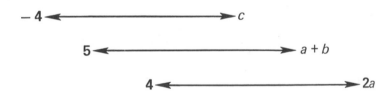

If $2a = 4$, then $a = 2$. If $a + b = 5$, then $2 + b = 5$, $b = 3$. $c = -4$. This verifies the expression $2n^2 + 3n - 4$. The common difference in the second subtraction is twice the n^2 coefficient.

We could have substituted other <u>corresponding values</u>. For example, using the next set of values on the table:

$$1 \longleftrightarrow a + b + c$$
$$9 \longleftrightarrow 3a + b$$
$$4 \longleftrightarrow 2a$$

If $2a = 4$, then $a = 2$. If $3a + b = 9$, then $6 + b = 9$, $b = 3$. And if $a + b + c = 1$, then $2 + 3 + c = 1$, so $c = -4$.

37

We get the same values of a, b, and c by this different substitution. As a matter of fact, we have usually started our tables with $n = 1$, not $n = 0$. Either one will do, but it is important to substitute corresponding values.

We have gone through how to find the general rule for any sequence which has a pattern that results in common differences after the second subtraction. Let's look at an example of how this is done with a problem from Section I of this book.

PROBLEM (p. 13)   Can you find the nth pentagonal number?

(1) Make a table of values and find the differences.

n	P		
1	1		
		>4	
2	5		>3
		>7	
3	12		>3
		>10	
4	22		>3
		>13	
5	35		

This shows that the general rule, or nth term, will be a <u>second</u> degree (quadratic) expression since the common difference came in the <u>second</u> subtraction.

(2) Substitute the values at the top of the table for the corresponding values in the general case. We have

$$1 \quad\quad\quad a+b+c$$
$$>4 \quad\quad\quad\quad >3a+b$$
$$>3 \quad\quad\quad\quad\quad\quad >2a$$

(3) Solve for a, b, and c.

$2a = 3$, $a = \frac{3}{2}$. $3a + b = 4$, $\frac{9}{2} + b = 4$, $b = -\frac{1}{2}$. $a + b + c = 1$, $\frac{3}{2} + (-\frac{1}{2}) + c = 1$, $c = 0$.

The rule is $\frac{3n^2}{2} - \frac{1}{2}n$ or $\frac{n(3n-1)}{2}$.

(4) Test your rule on terms in the sequence.

Notice that we substituted the value of $n = 1$, not $n = 0$, from the general value table.

n			
1	1		
		>4	
			>3

n			
0	c		
		$>a+b$	
1	$a+b+c$		$>2a$
		$>3a+b$	
			$>2a$

We did not record the value for $n = 0$ in our table of pentagonal numbers.

38

CAUTION

In using this method to solve problems, be careful not to equate values of $n = 1$ in your sequences with values of $n = 0$ in the general case. This error is frequently made.

Now we know how to solve problems which have a common difference in the second subtraction. What about those which have a common difference in the first, third, etc. subtraction?

The method is the same, as you might suspect. The following patterns hold.

common difference in		degree of expression		general form
first subtraction	➡	first degree expression	➡	$an + b$
second subtraction	➡	second degree expression	➡	$an^2 + bn + c$
third subtraction	➡	third degree expression	➡	$an^3 + bn^2 + cn + d$
fourth subtraction	➡	fourth degree expression	➡	$an^4 + bn^3 + cn^2 + dn + e$

Here are some examples of problems that do not have second degree rules.

PROBLEM   Find the nth term in the following sequence: 4, 11, 18, 25, . . .

n	T	
1	4	>7
2	11	>7
3	18	>7
4	25	

n	$an + b$	
1	$a + b$	$>a$
2	$2a + b$	$>a$
3	$3a + b$	$>a$
4	$4a + b$	

So, $a = 7$, $a + b = 4$, $7 + b = 4$, $b = -3$. The nth term is $7n - 3$.

An example of a problem with a common difference in the third subtraction follows.

PROBLEM   Find the nth term in the sequence -2, 0, 12, 40, 90, 168, . . .

n	T
1	-2
	> 2
2	0 > 10
	> 12 > 6
3	12 > 16
	> 28 > 6
4	40 > 22
	> 50 > 6
5	90 > 28
	> 78
6	168
$\vdots$	
n	___

n	$an^3 + bn^2 + cn + d$
1	$a + b + c + d$
	$> 7a + 3b + c$
2	$8a + 4b + 2c + d$ $> 12a + 2b$
	$> 19a + 5b + c$ $> 6a$
3	$27a + 9b + 3c + d$ $> 18a + 2b$
	$> 37a + 7b + c$ $> 6a$
4	$64a + 16b + 4c + d$ $> 24a + 2b$
	$> 61a + 9b + c$
5	$125a + 25b + 5c + d$
$\vdots$	
n	$an^3 + bn^2 + cn + d$

$6a = 6$, $a = 1$. $12a + 2b = 10$, $12 + 2b = 10$, $b = -1$. $7a + 3b + c = 2$, $7 - 3 + c = 2$, $c = -2$. $a + b + c + d = -2$, $1 - 1 - 2 = -2$, $d = 0$.

Therefore the nth term is $n^3 - n^2 - 2n$ or $n(n + 1)(n - 2)$ if factored.

SUMMARY

A generalization can be derived for the nth term of a sequence if there is a sufficient number of terms in the sequence to find a pattern of common differences between the terms or after a finite number of subtractions of the differences of these terms.

SOME LIMITATIONS

It is important to understand that a number of terms in a sequence may not <u>uniquely</u> determine a formula. Two criteria must be met to insure a unique generalization:

1. There must be a sufficient number of terms in the sequence to establish a pattern of common differences, and

2. it must be assumed that that pattern continues infinitely.

Examples of sequences including too few terms to assure a unique generalization are shown below.

Given the problem 1, 2, 4, ___, ___, . . ., here are three possible solutions:

1, 2, 4, <u>8</u>, <u>16</u>, . . . nth term: 2^{n-1}

or

1, 2, 4, <u>7</u>, <u>11</u>, . . . nth term: $\dfrac{n^2-n+2}{2}$

or

1, 2, 4, <u>8</u>, <u>15</u>, . . . nth term: $\dfrac{n(n^2-3n+8)}{6}$

Similarly,

1, 3, 6, <u>10</u>, <u>15</u>, <u>21</u>, . . . nth term: $\dfrac{n(n+1)}{2}$

1, 3, 6, <u>11</u>, <u>19</u>, <u>31</u>, . . . nth term: $\dfrac{(n^3-3n^2+14n-6)}{6}$

or

2, 4, 8, <u>14</u>, <u>22</u>, . . . nth term: n^2-n+2

2, 4, 8, <u>15</u>, <u>26</u>, . . . nth term: $\dfrac{(n+1)(n^2-n+6)}{6}$

2, 4, 8, <u>16</u>, <u>32</u>, . . . nth term: $2n$

It should be clear that in sequences like the examples shown above, more terms are needed to <u>establish</u> <u>a</u> <u>pattern</u> <u>of</u> <u>common</u> <u>differences</u> which would make the next missing terms unique.

A general proof for the technique of finite differences is not included here. A proof of the method is given by Betty L. Baker in the April 1967 issue of *School Science & Mathematics Magazine* (see bibliography). For another detailed explanation of sequences and series in general, refer to the *USSR Olympiad Problem Book* (see bibliography).

An interesting approach to proving generalizations about certain kinds of sequences and series is called mathematical induction. A brief explanation of this technique is given in the section on supplementary material on page 113 .

PROBLEM SECTION II

"The idea that aptitude for mathematics is rarer than aptitude for other subjects is merely an illusion which is caused by belated or neglected beginners."

Herbart

■ **What's My Rule?**

■ **What's My Sum?**

■ **Cannonball Problems**

■ **Regions in a Circle**

■ **Triangles in a Triangle**

■ **Squares in a Square**

■ **Squares on a Geoboard**

■ **Ten Men in a Boat**

■ **Tower of Hanoi**

■ **Shifting Pennies Puzzle**

WHAT'S MY RULE?

For each of the following sequences write the next three terms and the nth term.

nth term

1 1, 4, 9, 16, 25, ___, ___, ___, . . . , _____

2 4, 7, 12, 19, 28, ___, ___, ___, . . . , _____

3 2, 6, 12, 20, 30, ___, ___, ___, . . . , _____

4 3, 12, 27, 48, 75, ___, ___, ___, . . . , _____

5 −1, 5, 15, 29, 47, ___, ___, ___, . . . , _____

6 3, 7, 13, 21, 31, ___, ___, ___, . . . , _____

7 2, 7, 16, 29, 46, ___, ___, ___, . . . , _____

8 3, 3, 5, 9, 15, ___, ___, ___, . . . , _____

9 2, 5, 9, 14, 20, ___, ___, ___, . . . , _____

10 0, $\frac{4}{3}$, 4, 8, $\frac{40}{3}$, ___, ___, ___, . . . , _____

44

WHAT'S MY SUM?

How many sums can you find for the series below? Express your answer as a general rule. Test your rule when $n = 1$, $n = 2$, etc.

1 $1 + 2 + 3 + 4 + 5 + \ldots + n =$ _____

2 $1 + 3 + 5 + 7 + 9 + \ldots + (2n - 1) =$ _____

3 $3 + 4 + 5 + 6 + 7 + \ldots + (n + 2) =$ _____

4 $2 + 6 + 10 + 14 + 18 + \ldots + (4n - 2) =$ _____

5 $5 + 8 + 11 + 14 + 17 + \ldots + (3n + 2) =$ _____

6 $1 + 5 + 9 + 13 + 17 + \ldots + (4n - 3) =$ _____

7 $4 + 9 + 14 + 19 + 24 + \ldots + (5n - 1) =$ _____

8 $11 + 21 + 31 + 41 + 51 + \ldots + (10n + 1) =$ _____

9 $-5 + (-1) + 3 + 7 + 11 + \ldots + (4n - 9) =$ _____

10 $0 + (-2) + (-4) + (-6) + (-8) + \ldots + (-2n + 2) =$ _____

11 $\frac{1}{2} + 1 + \frac{3}{2} + 2 + \frac{5}{2} + \ldots + \frac{n}{2} =$ _____

45

CANNONBALL PROBLEMS

During the Civil War, the men in Company A stacked their cannonballs in a *triangular pyramid*.

1 How many cannonballs were there in the fourth layer below the top of the stack? In the tenth layer? In the nth layer?

2 How many cannonballs were stacked in the top ten layers?

3 Find a rule to determine the total number of cannonballs in any given number of layers, counting from the top. (i.e. Find a rule for the nth triangular pyramidal number.)

The men in Company B preferred to stack their cannonballs in a *square pyramid.*

4 How many cannonballs were there in the fifth layer below the top of the stack? In the ninth layer?

5 How many cannonballs were stacked in the top nine layers?

6 Find a rule to determine the total number of cannonballs in any given number of layers, counting from the top. (i.e. Find a rule for the nth square pyramidal number.)

46

REGIONS IN A CIRCLE

A *chord* is a line segment joining two points on a circle. Here *n* is the number of chords.

$n = 0$

$n = 1$

$n = 2$

$n = 3$

In the table to the right, R represents the maximum number of regions formed by a given number of chords in a circle.

Complete the table at the right through eight chords.

Find a pattern which will enable you to find the maximum number of regions formed by 60 chords without drawing or counting them.

How did you find the pattern?

Express the regions formed by n chords as a general rule.

n	R
0	
1	
2	
3	
4	
5	
6	
7	
8	
⋮	
60	
⋮	
n	

TRIANGLES IN A TRIANGLE

One *equilateral triangle* can be divided into many smaller equilateral triangles. In this problem, count only point-up triangles but count point-up triangles of all sizes.

DON'T COUNT point-down triangles.

DO COUNT point-up triangles made of other triangles.

  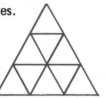 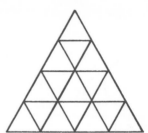

In the table to the right, n represents the number of units on a side. T represents the number of different equilateral triangles (apex-up), of all sizes, contained in a given equilateral triangle.

Complete the table at the right through seven units on a side.

Find a pattern which will enable you to find the general rule for the number of point-up equilateral triangles found in an n-sided equilateral triangle.

How did you find the pattern?

Express the number of equilateral triangles in an n-sided equilateral triangle as a general rule.

Counting *all* equilateral triangles in an n-sided triangle is a very difficult problem. You may wish to try it sometime.

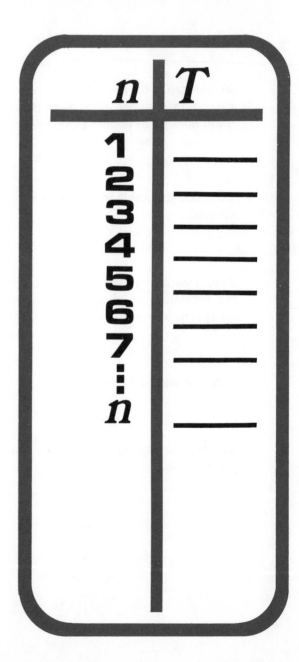

n	T
1	
2	
3	
4	
5	
6	
7	
...n	

SQUARES IN A SQUARE

A large square can be divided into many smaller squares. In this problem, be sure to count __all__ squares but do not count rectangles that are not square. Here, n represents the number of units on a side of the large square.

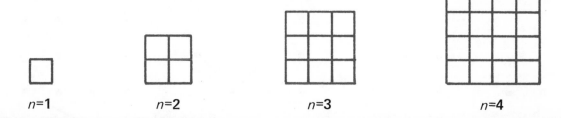

n=1   n=2   n=3   n=4

In the table to the right, S represents the number of squares of all sizes that are contained in a given square.

Complete the table at the right through a nine by nine square.

Find a pattern in the table which will enable you to find the number of squares in a given square.

How did you find the pattern?

Express the number of squares in an n x n square as a general rule.

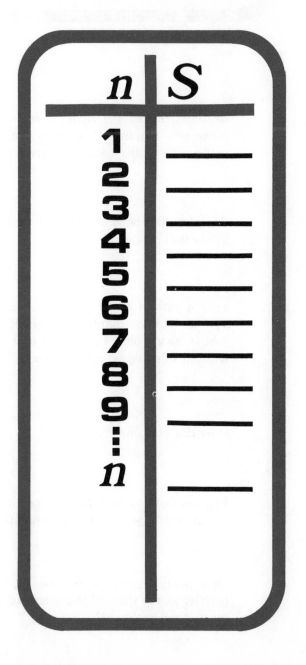

n	S
1	
2	
3	
4	
5	
6	
7	
8	
9	
...	
n	

49

SQUARES ON A GEOBOARD

Rubber bands can divide a *geoboard* into many squares. Some squares have sides parallel to the sides of the geoboard. Some squares have sides that are not parallel to the sides of the geoboard. Here, n represents the number of units on an edge. (An n x n geoboard contains $(n + 1)^2$ pins.)

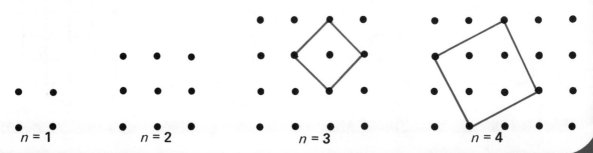

$n = 1$ $n = 2$ $n = 3$ $n = 4$

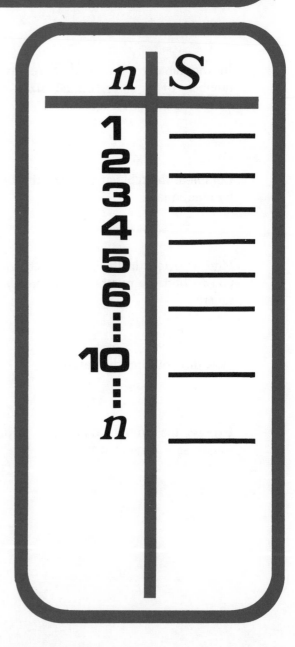

n	S
1	___
2	___
3	___
4	___
5	___
6	___
⋮	___
10	___
⋮	___
n	___

In the table to the right, S represents the maximum number of squares that can be made with rubber bands on a given geoboard.

Complete the table at the right through a six by six unit geoboard.

Find a pattern in the table which will enable you to find the number of squares made by rubber bands on a ten by ten unit geoboard without drawing or counting them.

How did you find the pattern?

Express how to find the number of squares on an n x n geoboard as a general rule.

How many squares with sides that are not parallel to the sides of the geoboard are there on an n x n geoboard?

50

TEN MEN IN A BOAT

Ten men are fishing in a boat. One seat in the center of the boat is empty. The five men in the front of the boat want to change seats and fish in the back of the boat and the five men in the back of the boat want to fish from the front of the boat. A man may move from his seat to the next empty seat, or he may step over one man without capsizing the boat. What is the minimum number of moves it will take to exchange the five men in front with the five in back?

Complete the table at the right for six pairs of men. Find a pattern that will enable you to find the number of moves it will take 12 pairs of men.

How did you find the pattern?

Express how to find the number of moves it will take n pairs of men as a general rule. What happens to the general rule if you consider it for the number of men in the boat rather than pairs of men?

Can you find a rule that works for odd and even numbers of men?

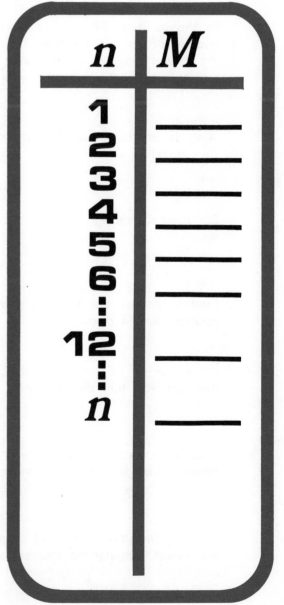

n	M
1	
2	
3	
4	
5	
6	
12	
n	

51

TOWER OF HANOI

The object of this ancient puzzle is to transfer the tower of discs to either of the two vacant pegs in the fewest possible moves. You may only move one disc at a time. You may not place a disc on one that is smaller.

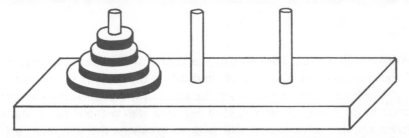

In the table to the right, *n* represents the number of discs in the tower. *M* represents the fewest number of moves it takes to transfer those discs to the vacant pegs.

What is the fewest number of moves with four discs?

Complete the table at the right through seven discs.

Find a pattern which would give you the solution for 64 discs.

How did you find the pattern?

What is the formula for the fewest number of moves needed to transfer *n* number of discs?

Find a pattern for the number of moves each disc makes. Consider the smallest disc number one.

n	M
1	
2	
3	
4	
5	
6	
7	
⋮	
64	
⋮	
n	

SHIFTING PENNIES PUZZLE

To invert a triangular arrangement of three pennies, the minimum number of pennies that must be moved is one.

By moving one coin, this arrangement is changed to this

As more rows of coins are added, the number of coins that must be moved is increased.

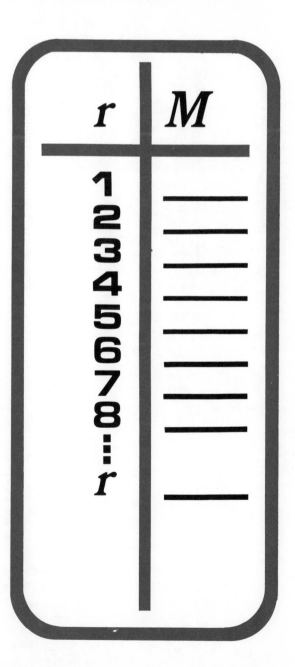

r represents the number of rows of coins in the triangular arrangement and M is the minimum number of coins that must be moved to invert the triangle. Find a pattern and try to develop expressions for M in terms of r. Hint: Consider every third row in the data you obtain.

PROBLEM SECTION III

"Mathematicians assume the right to choose, within the limits of logical contradiction, what path they please in reaching their results."

Henry Adams

■ What's My Rule?

■ What's My Sum?

■ Rectangles in a Square

■ Red-Faced Cubes

■ Mosaics

■ Variation of the Method

■ Fences and Corrals

■ Polygonal Numbers

■ Comparing Series

■ Series Synthesis

■ Pyramidal Numbers

■ Pythagorean Triples

■ Pick's Theorem

■ Dot Paper

■ Grasshopper Games

■ Game Sheet

■ Circles

■ Networks

■ Network Diagram

■ Multipaths

WHAT'S MY RULE?

Complete the sequences.

1   8,   27,   64,   125,   216,   ___,   ___,   . . . ,   _____

2   1,   8,   27,   64,   125,   ___,   ___,   . . . ,   _____

3   3,   11,   31,   69,   131,   ___,   ___,   . . . ,   _____

4   4,   19,   44,   79,   124,   ___,   ___,   . . . ,   _____

5   −3,   11,   49,   123,   245,   ___,   ___,   . . . ,   _____

6   5,   11,   19,   29,   41,   ___,   ___,   . . . ,   _____

7   −1,   0,   1,   8,   27,   ___,   ___,   . . . ,   _____

8   −2,   6,   26,   64,   126,   ___,   ___,   . . . ,   _____

9   7,   25,   59,   115,   199,   ___,   ___,   . . . ,   _____

10   6,   24,   60,   120,   210,   ___,   ___,   . . . ,   _____

WHAT'S MY SUM?

How many sums can you find for the series below? Express your answer in terms of n.

1 $1 + 3 + 6 + 10 + 15 + \ldots + \frac{1}{2}n(n+1) =$ _____

2 $1 + 4 + 9 + 16 + 25 + \ldots + n^2 =$ _____

3 $1 + 5 + 12 + 22 + 35 + \ldots + \frac{1}{2}n(3n-1) =$ _____

4 $1 + 6 + 15 + 28 + 45 + \ldots + n(2n-1) =$ _____

5 $3 + 7 + 13 + 21 + 31 + \ldots + (n^2 + n + 1) =$ _____

6 $1 \cdot 2 + 2 \cdot 3 + 3 \cdot 4 + 4 \cdot 5 + 5 \cdot 6 + \ldots + n(n+1) =$ _____

7 $13 + 29 + 51 + 79 + 113 + \ldots + (3n^2 + 7n + 3) =$ _____

8 $\frac{3}{2} + 4 + \frac{15}{2} + 12 + \frac{35}{2} + \ldots + \frac{1}{2}n(n+2) =$ _____

9 $1 + 8 + 27 + 64 + 125 + \ldots + n^3 =$ _____

10 $1 + 27 + 125 + 343 + 729 + \ldots + (2n-1)^3 =$ _____

WHAT'S MY SUM?

How many sums can you find for the series below? Express your answer in terms of n.

1 $1^2 + 3^2 + 5^2 + 7^2 + 9^2 + \ldots + (2n-1)^2 = \underline{\hspace{2cm}}$

2 $2^2 + 4^2 + 6^2 + 8^2 + 10^2 + \ldots + (2n)^2 = \underline{\hspace{2cm}}$

3 $1^4 + 2^4 + 3^4 + 4^4 + 5^4 + \ldots + n^4 = \underline{\hspace{2cm}}$

4 $1 \cdot 3 + 2 \cdot 4 + 3 \cdot 5 + 4 \cdot 6 + 5 \cdot 7 + \ldots + n(n+2) = \underline{\hspace{2cm}}$

5 $2 \cdot 3 + 4 \cdot 5 + 6 \cdot 7 + 8 \cdot 9 + 10 \cdot 11 + \ldots + 2n(2n+1) = \underline{\hspace{2cm}}$

6 $\frac{1}{3} + 1 + \frac{5}{3} + \frac{7}{3} + 3 + \ldots + \frac{2n-1}{3} = \underline{\hspace{2cm}}$

7 $1 \cdot 2 \cdot 3 + 2 \cdot 3 \cdot 4 + 3 \cdot 4 \cdot 5 + 4 \cdot 5 \cdot 6 + 5 \cdot 6 \cdot 7 + \ldots + n(n+1)(n+2) = \underline{\hspace{2cm}}$

8 $1 + 2 + 3 + 4 + \ldots + n + \ldots + 4 + 3 + 2 + 1 = \underline{\hspace{2cm}}$

9 $\frac{1}{1 \cdot 2} + \frac{1}{2 \cdot 3} + \frac{1}{3 \cdot 4} + \frac{1}{4 \cdot 5} + \frac{1}{5 \cdot 6} + \ldots + \frac{1}{n(n+1)} = \underline{\hspace{2cm}}$

10 $\frac{1}{1 \cdot 3} + \frac{1}{3 \cdot 5} + \frac{1}{5 \cdot 7} + \frac{1}{7 \cdot 9} + \frac{1}{9 \cdot 11} + \ldots + \frac{1}{(2n-1)(2n+1)} = \underline{\hspace{2cm}}$

RECTANGLES IN A SQUARE

A square can be divided into many rectangles. Two rectangles are considered different if they do not have the same four vertices. Count __all__ different rectangles. Here, n represents the number of rectangles on a side of a square.

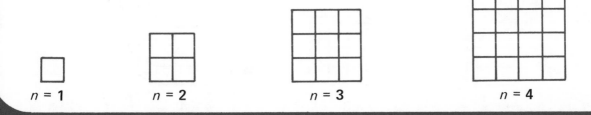

$n = 1$ $n = 2$ $n = 3$ $n = 4$

In the table to the right, R represents the number of rectangles in a given square.

Complete the table at the right through eight units on the side of a square.

Find a pattern which will enable you to find the number of different rectangles in a 25 x 25 square without drawing or counting them.

How did you find the pattern?

Express the formula for the number of rectangles in an n x n square as a general rule.

The general expression for the nth term is the same as the expression for the sum of what special series? See page 57.

n	R
1	____
2	____
3	____
4	____
5	____
6	____
7	____
8	____
⋮	
25	____
⋮	
n	____

59

RED-FACED CUBES

The faces of a cube that measures two units on an edge are all red. If the cube is cut into unit cubes, there will be eight cubes. Each of the eight will have three red faces. Repeating the process with cubes measuring three or more units on an edge will result in unit cubes having three, two, one, or no red faces.

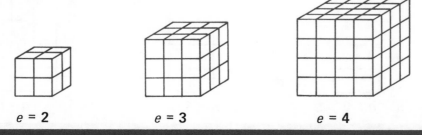

$e = 2$ $e = 3$ $e = 4$

Find a pattern and write an expression for the number of each kind of unit cube obtained in terms of the length of the edge of the original cube.

Complete the chart below.

length of edge e	number of unit cubes	three red faces	two red faces	one red face	no red faces
2	_____	_____	_____	_____	_____
3	_____	_____	_____	_____	_____
4	_____	_____	_____	_____	_____
5	_____	_____	_____	_____	_____
6	_____	_____	_____	_____	_____
7	_____	_____	_____	_____	_____
⋮					
e	_____	_____	_____	_____	_____

To check, show that the sum of the general terms in the four columns on the right equals e^3.

60

MOSAICS

Bees are clever engineers. The cross section of a honeycomb is a mosaic of hexagons with each corner surrounded by three hexagons. See the lower right figure.

Is this the most efficient design for storage of honey? To find out, let's compare two cells, one with a hexagonal base and the other with a square base. If these two cells have the same height and their bases have the same perimeter, then they would require the same amount of wax to build. The volume of a *prism* (both cells are prisms) is the area of the base times height. Therefore, for a given height, the cell with the greater volume is the one whose base has the greater area.

1 What is the area of a square which has a perimeter of 12?

2 What is the area of a hexagon which has a perimeter of 12?

3 Which of the polygons in problems 1 and 2 has the greater area? For a given height, which cell (square or hexagonal) has the greater volume?

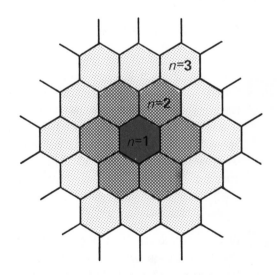

4 Which would be the best design, a cell with four, six, eight, ten, or 12 sides? Why?

5 This is a two-dimensional problem: Start with one hexagon ($n=1$) and surround it with six hexagons ($n=2$). Then surround these with more hexagons ($n=3$), etc. Find an expression for the total number of hexagons in this pattern in terms of n.
Hint: $S_n = 1 + 6 + 12 +$ ___ $+$ ___ $+$ ___ ...

MOSAICS

6 Will the expression that you obtained in problem 5 always yield a prime number? If not, what is an exception?

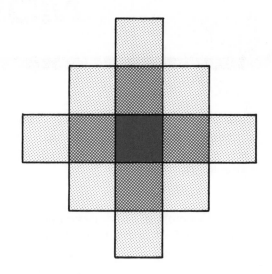

7 Start with one square ($n=1$) and surround it with four squares ($n=2$). Then surround these with more squares ($n=3$), etc. Find an expression for the total number of squares in terms of n. $S_n = 1 + 4 + 8 + 12 + 16 + \ldots$

8 When $n=4$, the 25 pieces of the mosiac could be rearranged to form a large square. Will this be possible again? If so, when? That is, will our sequence of mosaic sums 1, 5, 13, 25, 41, . . ., ever again contain a perfect square?

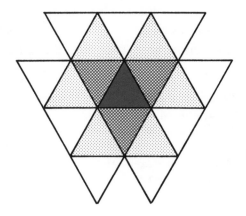

9 Start with one equilateral triangle ($n=1$) and surround it with three triangles ($n=2$). Then surround these with more triangles ($n=3$), etc. Find an expression for the total number of triangles in terms of n.
$S_n = 1 + 3 + 6 + 9 + 12 + \ldots$

10 When $n=3$, the mosaic above could be rearranged to form the pattern of apex-up triangles shown below. Will this be possible again? If so, when? That is, will our sequence of mosaic sums 1, 4, 10, 19, 31, . . . ever again contain any of the triangular numbers?

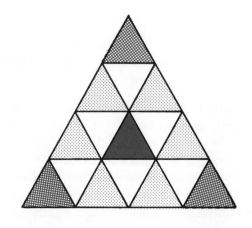

11 Write a sequence for the geometric design on the back cover with terms corresponding to the number of petals in the central figure and in each surrounding ring. What is the general term? Find a rule for determining the total number of petals in the design if there are n concentric rings of design.

62

VARIATION OF THE METHOD

For certain kinds of problems, a variation of the method of finite differences may be helpful. An example is shown below.

Sample relation

x	y
2	1
4	7
6	17
8	31
10	49

$$1 \overset{>6}{\underset{>10}{\quad}} \overset{>4}{\quad}$$

with first differences 6, 10, 14, 18 and second differences 4, 4, 4

General quadratic

x	$y = ax^2 + bx + c$
2	$4a + 2b + c$
4	$16a + 4b + c$
6	$36a + 6b + c$
8	$64a + 8b + c$

The relation in the example is quadratic since the second difference is constant, but the values of x are <u>not consecutive</u> natural numbers. So, using the general equation $y = ax^2 + bx + c$ and any three pairs of values from the table, three equations with three unknowns are written. Then these are solved simultaneously as shown below and a, b, and c in the general equation are replaced with values obtained.

(1)　$x = 6, y = 17$　　　　　　$36a + 6b + c = 17$

(2)　$x = 4, y = 7$　　　　　　$\underline{16a + 4b + c = 7}$

(3)　Subtract the step (2) equation from the step (1) equation.

　　　　　　　　　　　$20a + 2b = 10$ or $10a + b = 5$

(4)　$x = 2, y = 1$　　　　　　$4a + 2b + c = 1$

(5)　Subtract the step (4) equation from the step (2) equation.

(6)　Subtract the step (5) equation from the step (3) equation.

　　　　　　　　　　　　　　　$a = \frac{1}{2}$

Then, by substitution in step (5): $6a + b = 3$; $3 + b = 3$; $b = 0$.

By substitution in step (4): $4a + 2b + c = 1$; $2 + 0 + c = 1$; $c = -1$.

Finally, our general rule for the sequence above is $y = \frac{1}{2}x^2 - 1$. This method will be helpful in solving problems concerning triangle fences and corrals.

TRIANGLE FENCES AND CORRALS

A triangle fence is used to build a triangle corral.

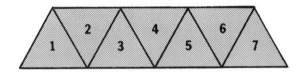

triangle fence

To make a corral, the triangle fence must be at least seven sections long. Why? Longer fences for corral sides require more triangles and increase the area inside the corral.

Use the drawing and the table on the next page to find the answers to these questions:

1 What is the relation between *S* and *F*?

2 What is the relation between *S* and *I*?

3 What is the relation between *S* and (*F+I*)? Use the results of questions 1 and 2 to verify this.

4 As *S* increases, will it ever be true that *I* > *F*? If so, for what value of *S*?

5 Will *I* and *F* ever be equal?

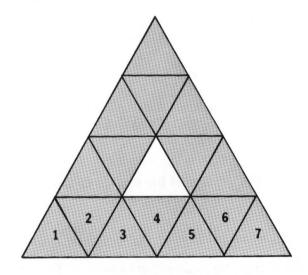

triangle corral

TRIANGLE FENCES AND CORRALS

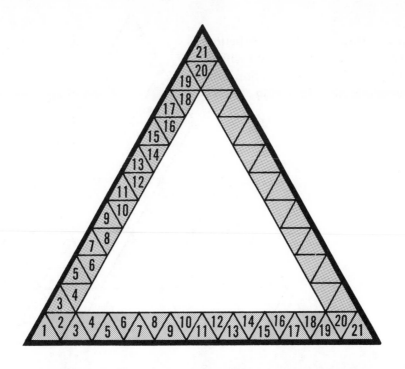

Number of triangles on one side of corral	Number of triangles in entire fence	Number of triangles inside fence	
S	F	I	F + I
7	—	—	—
9	—	—	—
11	—	—	—
13	—	—	—
15	—	—	—
⋮	⋮	⋮	⋮
n	—	—	—

TRIANGLE FENCES AND CORRALS

6 ($F + I$) is always a square number. To figure out why, study the triangle at the right which has the same shape as the grid on the previous page, and complete the following table.

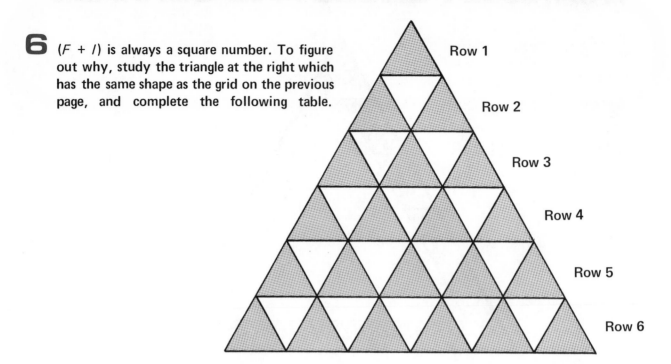

Row 1
Row 2
Row 3
Row 4
Row 5
Row 6

Row	Triangles in each row			Triangles on or above each row		
	Red	White	Total	Red	White	Total
Row 1	___	___	___	___	___	___
Row 2	___	___	___	___	___	___
Row 3	___	___	___	___	___	___
Row 4	___	___	___	___	___	___
Row 5	___	___	___	___	___	___
Row 6	___	___	___	___	___	___

7 Observation: The sum of two consecutive _____ numbers is a _____ number. (See page 68 for more on this.)

8 Would it ever be possible to rearrange the red and white triangles and obtain a triangle corral with red triangles forming the fence surrounding all the white triangles? That is, will F and I ever be consecutive triangle numbers? If so, when?

POLYGONAL NUMBERS

Can you recall how the sequences of *polygonal numbers* start? Complete the table below finding a pattern, a generalization for each sequence in terms of n, and a generalization for the entire set in terms of S and n. If you need a hint, study the network diagram on page 83 rather than referring to earlier problems and solutions.

	t_1	t_2	t_3	t_4	t_5		t_n
Triangular $S = 3$	___,	___,	___,	___,	___,	⋯	___
Square $S = 4$	___,	___,	___,	___,	___,	⋯	___
Pentagonal $S = 5$	___,	___,	___,	___,	___,	⋯	___
Hexagonal $S = 6$	___,	___,	___,	___,	___,	⋯	___
Heptagonal $S = 7$	___,	___,	___,	___,	___,	⋯	___
Octagonal $S = 8$	___,	___,	___,	___,	___,	⋯	___
Nonagonal $S = 9$	___,	___,	___,	___,	___,	⋯	___
Decagonal $S = 10$	___,	___,	___,	___,	___,	⋯	___
$\vdots$							
S	___,	___,	___,	___,	___,	⋯	___

COMPARING SERIES

Many interesting patterns may be discovered by studying the completed table of polygonal numbers on page 67 and by adding or subtracting the corresponding terms of two different sequences or series. Three examples are given below.

Example I

Starting with the second term (t_2) the differences between the square and the triangular number series are triangular numbers.

square numbers	$1 + 4 + 9 + 16 + 25 \ldots$
triangular numbers	$1 + 3 + 6 + 10 + 15 \ldots$
difference	$0 + 1 + 3 + \ 6 + 10 \ldots$

Adding the second and third difference (1 and 3) we arrive at the second square number (4). Adding the third and fourth difference (3 and 6) we arrive at the third square number (9). This relationship can also be shown geometrically.

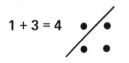

$$1 + 3 = 4 \qquad\qquad 3 + 6 = 9$$

In general, $\triangle_n + \triangle_{n+1} = \square_{n+1}$

1) Is it true that $\triangle_n + \pentagon_{n+1} = \hexagon_{n+1}$?

2) $3\triangle_n + \triangle_{n+1} =$ _____

Example II

Square Mosaic Numbers	$1 + 4 + 8 + 12 + 16 + \ldots$	$=$ _____
Odd Numbers	$1 + 3 + 5 + \ 7 + \ 9 + \ldots + (2n-1)$	$= n^2$
Differences	$0 + 1 + 3 + \ 5 + \ 7 + \ldots$	$=$ _____

This new series contains odd numbers, but only $(n-1)$ of them, so its sum is $n^2 - (2n-1)$ or $n^2 - 2n + 1$.

COMPARING SERIES

3) Use this and the sum of the odd numbers to find the sum of the Square Mosaic Numbers. Does this check with the results you obtained on page 62?

Example III

Cubic Numbers	1,	8,	27,	64,	125,	216,	... n^3
Square Pyramidal Numbers	1,	5,	14,	30,	55,	91,	... ____
Differences	0,	3,	13,	34,	70,	125,	... ____

4) Find the general term of this new sequence. Check it by subtracting general terms of the original sequences.

There are many other possibilities for comparison using the sequences and series that you have already studied and there will be more after you have completed all the problems in this book. You may wish to make some comparisons on your own. If so, your investigations will increase your understanding. Some of the results may surprise you.

SERIES SYNTHESIS

Nearly all of the problems in this book have involved analyzing mathematical sequences and series. Now let's consider the opposite, building some sequences and series. We will do this using repeated addition.

Other kinds of sequences and series may be formed by using different operations. For example, repeated multiplications generate *geometric sequences* and series, a related topic which is not included in this book except for the Tower of Hanoi problem and problems 9 and 10 on page 57.

Here is an example of the steps which generate a sequence. Let the first term equal 1 (a=1) and the common difference equal 3(d=3).

(1) $1 + 3 + 3 + 3 + 3 + ...$
(2) $1, 1 + 3, 1 + 3 + 3, 1 + 3 + 3 + 3, 1 + 3 + 3 + 3 + 3, ...$ or $1, 4, 7, 10, 13, ...$
(3) $1 + 4 + 7 + 10 + 13 + ...$
(4) $1, 5, 12, 22, 35, ...$
(5) $1 + 5 + 12 + 22 + 35 + ...$
(6) $1, 6, 18, 40, 75, ...$

The steps involved in generating a sequence or series are explained below.

(1)   Select numbers for the first term and for the common differences and write a series using these numbers.

(2)   Write the sequence of sums as the number of terms in the series increases. For instance, $t_1, t_1 + t_2, t_1 + t_2 + t_3$, etc.

(3)   Insert plus signs between the terms of the sequence to change it to a series.

(4)   Repeat step (2).

(5)   Repeat step (3).

(6)   Repeat step (2).

SERIES SYNTHESIS

A rule frequently used to find an expression for the nth, or general, term of an *arithmetic sequence* (see step (2) page 70) is $t_n = a + (n-1)d$. You may see how this is obtained by noting that if there are n terms in step (1), then $(n-1)$ of them are 3's.

1 What is the general term in step (2) on page 70?

Another formula is frequently used to find the sum of n terms of an arithmetic series (see step (3) on page 70). It is $S_n = \frac{n}{2}(a+l)$, where l represents the last term to be included in the sum and can be found using the formula for t_n (see page 23, Gauss' approach).

2 What is the sum of n terms in step (3) on page 70?

3 There are polygonal numbers in step (4) on page 70. Which are they?

4 What is the general term in step (4) on page 70?

5 What is the sum in step (5) on page 70?

6 The numbers in step (6) on page 70 are called pentagonal pyramidal numbers. What is the general term?

PYRAMIDAL NUMBERS

Pyramidal numbers are numbers that can be represented by objects in a pyramidal array.

Complete the table below finding a generalization for each sequence in terms of n and one for the entire set in terms of S and n.

	t_1	t_2	t_3	t_4	t_5		t_n
Triangular $S = 3$	_____,	_____,	_____,	_____,	_____,	⋯	_____
Square $S = 4$	_____,	_____,	_____,	_____,	_____,	⋯	_____
Pentagonal $S = 5$	_____,	_____,	_____,	_____,	_____,	⋯	_____
Hexagonal $S = 6$	_____,	_____,	_____,	_____,	_____,	⋯	_____
Heptagonal $S = 7$	_____,	_____,	_____,	_____,	_____,	⋯	_____
Octagonal $S = 8$	_____,	_____,	_____,	_____,	_____,	⋯	_____
Nonagonal $S = 9$	_____,	_____,	_____,	_____,	_____,	⋯	_____
Decagonal $S = 10$	_____,	_____,	_____,	_____,	_____,	⋯	_____
⋮ S	_____,	_____,	_____,	_____,	_____,	⋯	_____

Note: If you use spheres, such as styrofoam balls, to make a model of pyramidal numbers, you may have difficulty. Keeping the bottom layer in the polygonal shape, for $S=3$ and $S=4$ the spheres would be closely packed allowing each layer to support the layers above. However, for $S \geqslant 5$ the spheres will not nest together neatly. The pyramid may slump due to the space between the spheres in the interior.

PYTHAGOREAN TRIPLES

A *Pythagorean number triple* is a set of three numbers, *a*, *b*, and *c* where $a^2 + b^2 = c^2$. If *a*, *b*, and *c* contain no common factors the triple is *primitive*.

Here are four sets of Pythagorean number triples. For each set, find the missing numbers and the general expression for the numbers in each column. Use the *Pythagorean Theorem* to check your results. Which of the triples in set 4 are not primitive?

1

n	a	b	c
1	1	0	1
2	3	4	5
3	5	12	13
4	7	24	25
5	9	40	41
6	11	60	61
7	13	84	85
⋮			
21	—	—	—
⋮			
n	—	—	—

2

n	a	b	c
1	4	3	5
2	8	15	17
3	12	35	37
4	16	63	65
5	20	99	101
⋮			
10	—	—	—
⋮			
n	—	—	—

3

n	a	b	c
1	12	5	13
2	20	21	29
3	28	45	53
4	36	77	85
5	44	117	125
⋮			
12	—	—	—
⋮			
n	—	—	—

4

n	a	b	c
1	15	8	17
2	21	20	29
3	27	36	45
4	33	56	65
5	39	80	89
6	45	108	117
7	51	140	149
8	57	176	185
9	63	216	225
10	69	260	269
⋮			
n	—	—	—

PICK'S THEOREM

If a polygon is constructed on a regular geoboard, a generalization can be made concerning the relationship between the area, A, of the figure, the number of points on the boundary, N_b, and the number of points in the interior, N_i. This generalization is called *Pick's Theorem.* It was first published by G. Pick in a Czechoslovakian journal in 1899.

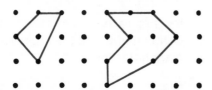

$N_b = 4$ $N_b = 8$
$N_i = 1$ $N_i = 2$
$A = 2$ $A = 5$

Use the tables below to find A in terms of N_b and N_i. The dot paper on the next page may be useful.

3 boundary points

N_i	A
0	____
1	____
2	____
3	____
4	____
.	
.	
.	
N_i	____

4 boundary points

N_i	A
0	____
1	____
2	____
3	____
4	____
.	
.	
.	
N_i	____

5 boundary points

N_i	A
0	____
1	____
2	____
3	____
4	____
.	
.	
.	
N_i	____

N_b boundary points

N_b	A
3	____
4	____
5	____
6	____
.	
.	
.	
N_b	____

DOT PAPER

GRASSHOPPER GAMES

GRASSHOPPER SOLITAIRE

This game is played on a nine by nine array of squares with nine playing pieces (pennies, buttons, etc.) arranged in a three by three array in one corner. The object is to move all of the pieces to a three by three arrangement in the opposite corner in the <u>minimum</u> number of moves. Moves may be made horizontally, vertically, diagonally, or by jumping in any direction over an adjacent piece. More than one jump may be made, but each jump counts as a move. Playing pieces jumped are not removed.

For a given number of playing pieces p, the minimum number of moves M depends on the length of a side of the playing square S. Using the playing squares on the next page and recording results, find expressions for M in terms of S for these variations of the game.

1 $p = 1$; $S = 2, 3, 4, 5$.

2 $p = 4$; $S = 3, 4, 5, \ldots$.

3 $p = 9$; values of S as needed.

4 $p = 16$; values of S as needed.

5 Are there any restrictions on the minimum value of S?

TWO-PLAYER GRASSHOPPER

In the two-player version of the game, players arrange playing pieces in identical arrays (triangular, square, etc.) in opposite corners of the nine by nine playing square. The players' pieces should be distinct. Players take turns moving. Moves are the same as in Grasshopper Solitaire. Jumping over the opponent's playing pieces is allowed. The first player to move his playing pieces to the original array in the diagonally opposite corner is the winner.

GRASSHOPPER GAME SHEET

CIRCLES

POINTS AND CHORDS OF CIRCLES

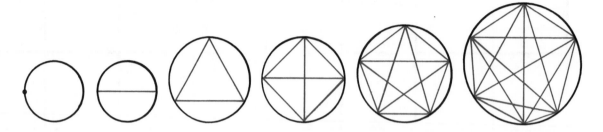

Figure 1

If there are two points on a circle, one chord may be drawn joining them. A third point on the circle makes three chords possible, and a fourth point makes six chords possible. The greater the number of points, the greater the number of chords that may be drawn.

1 Find the relation between the number of points P and the number of chords C.

2 Do you recognize this polynomial from previous problems? If so, which?

CHORDS AND REGIONS OF A CIRCLE

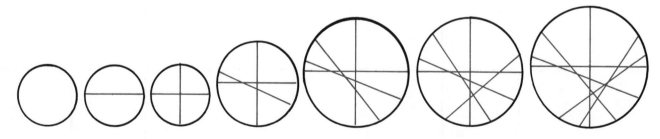

Figure 2

One chord separates the interior of a circle into two regions. Two chords that intersect in the interior of a circle separate it into four regions. Three chords that are not *concurrent* separate it into a maximum of seven regions.

3 Find the relation between the number of chords C and the maximum number of regions R.

CIRCLES

4 You may wish to draw other figures and investigate the relation between C and R for chords that are drawn in different ways, such as:

a. chords that are parallel;

b. chords that have a common end point;

c. chords that are all concurrent at a point within the circle;

d. chords drawn so that each intersects two others at a point within the circle. For example, the second chord intersects the first, the third intersects the second, and so on until the nth chord intersects the $(n-1)$th and also the first chord.

CHORDS AND POINTS OF INTERSECTION

See figure 2.

If two chords intersect within a circle, then a third chord may be drawn which will intersect each of the chords at a different point, giving a total of three points of intersection. If each new chord intersects each of the other chords, and no three chords are concurrent, then the number of points of intersection will be maximum.

5 Find the relation between the number of chords C and the maximum number of points of intersection I.

6 Is this polynomial the same as the result of any other problems? If so, which?

7 If two chords intersect within a circle, the point of intersection separates each chord into two segments. Using figure 2, page 78, find the relation between the number of chords C and the number of segments S.

8 Using the results of problems 5 and 7, find a relation involving C, I, and S. You may wish to test the relation involving C, I, and S on figure 1 page 78, and figures you may have drawn for problem 4 to determine if there are any restrictions regarding the number of chords intersecting at a point and the location of the point of intersection.

CIRCLES

POINTS ON A CIRCLE AND INTERSECTION
OF CHORDS

9 Use figure 1 on page 78 and the figure below for this problem. Find the relation between *P*, the number of points on the circle and *I*, the number of points formed by chords intersecting within the circle, where *I* is to be a maximum. The chords represented by black dotted lines, forming the sides of an *inscribed polygon*, should be disregarded for this problem.

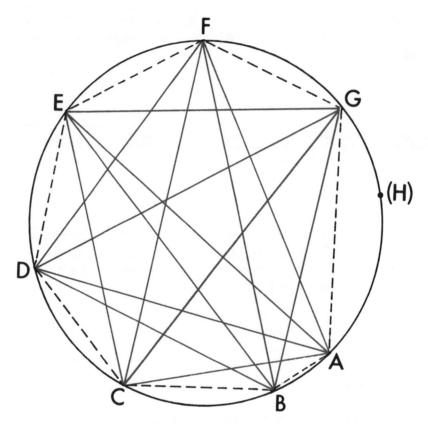

10 An interesting pattern of intersections may be found by starting with a given number of points on a circle and drawing chords connecting non-adjacent points. In the example above, we draw four chords from point A, four chords from point B, three chords from point C, two chords from point D, etc. Note the number of intersections along the chords from the seven points taken in order. Can you find the pattern? Try introducing an eighth point, like (H), on the circle to confirm your *conjecture*.

CIRCLES

POINTS AND REGIONS OF A CIRCLE

11 Using figure 1 on page 78 and the figure on page 80, find the relation between P, the number of points on a circle, and R, number of non-overlapping regions formed within the circle, where R is to be a maximum.

Warning: Check the data carefully for $P=6$ and $P=7$. Do not jump to a premature conclusion.

REGIONS FORMED BY INTERSECTING CIRCLES

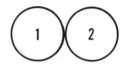
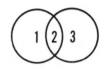

A single circle encloses one region. Two circles enclose one, two, or a maximum of three regions depending on whether they coincide, are tangent, or intersect at two points. The greater the number of circles, the greater the number of regions they can be made to enclose.

12 Using the values of one through six for n, find the relation between the number of circles n, and the maximum number of regions R, enclosed by the circles. By drawing some intersecting circles and recording data, you should be able to answer the questions below. If you have difficulty drawing figures, you may want to cut out rings and arrange them.

(Note: If the circles are congruent and the arrangement is symmetrical, you may discover that points of intersection determine vertices of regular polygons which are concentric and related to the number of circles.)

13 Let R' represent the maximum number of regions into which the plane may be separated by n intersecting circles and find the relation between n and R'.

NETWORKS

In the 18th century, a Swiss mathematician, Leonard Euler (oil-er) discovered a rule for a simple *polyhedron*: the number of faces plus the number of vertices, minus the number of edges, equals two.
$$F + V - E = 2$$

Examples:

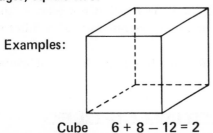

Cube $6 + 8 - 12 = 2$

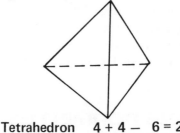

Tetrahedron $4 + 4 - 6 = 2$

You may wish to verify this rule for other three-dimensional figures such as a square-based pyramid, a hexagonal prism, the other regular polyhedra (*octahedron, dodecahedron, icosahedron,* etc.).

There is a similar rule for two-dimensional figures or *networks*. It may be stated in terms of R, P, and L where R represents the number of separate enclosed regions, P represents the total number of points in the network, and L represents the number of non-overlapping line segments joining points of the network.

1 Examine the figures on page 83 and record the values of R, P, and L. Can you discover this rule for networks?

2 If n represents the number of points on a side of the triangle, find the relations between n and R, n and P, n and L. Use the data recorded in problem 1, and the expressions obtained, to verify the network rule.

3 Repeat problem 2 letting n represent the number of points on a side of the square.

4 Repeat problem 2 letting n represent the number of points on a side of the pentagon.

5 Repeat problem 2 letting n represent the number of points on a side of the hexagon.

6 Can the network rule be applied to one-dimensional figures, i.e. to points on a line?

7 You may wish to test this rule on networks of intersecting circles, where L represents the number of arcs joining adjacent points of intersection. See problem 12 on page 81, where the maximum number of regions R was expressed in terms of the number of circles n. Find relations between n and P, and n and L and use these expressions to verify the network rule.

REGULAR POLYGONAL NETWORKS

$n = 1$ $n = 2$ $n = 3$ $n = 4$ $n = 5$ $n = 6$

   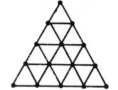 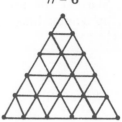

    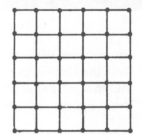

  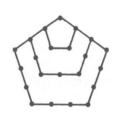 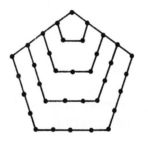 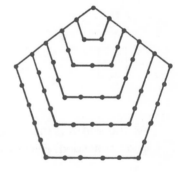

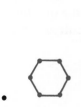 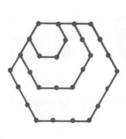 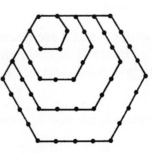 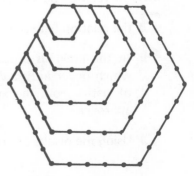

MULTIPATHS

Start at the lower left corner of each rectangular grid and move along the vertical and horizontal lines toward the upper right corner. You can only move up or to the right.

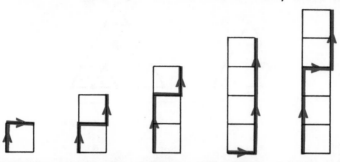

How many different paths are there? The larger the grid, the greater the number of paths and the more important a systematic approach becomes. One approach would be to find the number of paths to the goal along the interior grid lines from points on the bottom and from points on the left side and then add these.

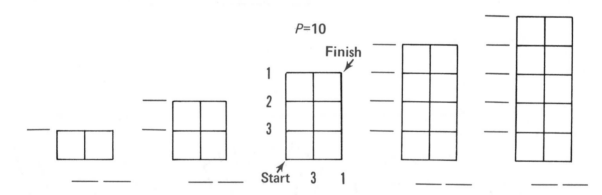

As the width is increased, the results obtained from narrower grids may be used. Draw more grids as needed. Record the data and find:

1 the relation between the number of paths P and the length ℓ, if the width of the grid equals 1 ($w=1$);

2 the relation between P and ℓ if $w=2$;

3 the relation between P and ℓ if $w=3$.

4 In the previous problem you may have discovered triangular numbers and other numbers which occur in *Pascal's Triangle.* Did you consider revising the approach to find the number of paths from the starting point to each point of the grid? What is the result of this approach?

5 Using the results of problem 4, can you find the relation between P and ℓ if $w=4$?

SOLUTIONS

SOLUTIONS

"I am not sure that it is of great value in life to know how many diagonals an n-sided figure has. It is the method rather than the result that is valuable."

W. W. Sawyer

SOLUTIONS

SECTION 1

Solutions for these problems, using the technique of finite differences, form the second part of this section.

p. 9 WHAT'S MY RULE?

1) $12, 14, 16, \ldots, 2n$
2) $11, 13, 15, \ldots, 2n - 1$
3) $18, 21, 24, \ldots, 3n$
4) $22, 25, 28, \ldots, 3n + 4$
5) $30, 35, 40, \ldots, 5n$
6) $27, 32, 37, \ldots, 5n - 3$
7) $4, \frac{9}{2}, 5, \ldots, \frac{1}{2}n + 1$
8) $1, \frac{3}{2}, 2, \ldots, \frac{1}{2}n - 2$
9) $57, 65, 73, \ldots, 8n + 9$
10) $79, 93, 107, \ldots, 14n - 5$

p. 10 SQUARE NUMBERS

$1, 4, 9, 16, 25, 36, 49, 64, \ldots$
$400, \ldots, 10000, \ldots$ nth term: n^2

p. 11 TRIANGULAR NUMBERS

$1, 3, 6, 10, 15, 21, 28, 36, 45, \ldots,$
$465, \ldots, 5050, \ldots$
nth term: $\frac{1}{2}n(n+1)$

p. 12 OBLONG NUMBERS

$2, 6, 12, 20, 30, 42, 56, 72, \ldots,$
$650, \ldots, 10100, \ldots$
nth term: $n(n+1)$

p. 13 PENTAGONAL NUMBERS

$1, 5, 12, 22, 35, 51, 70, 92, \ldots,$
$145, \ldots, 14950, \ldots$
nth term: $\frac{1}{2}n(3n-1)$

p. 14 HEXAGONAL NUMBERS

$1, 6, 15, 28, 45, 66, 91, \ldots, 435,$
$\ldots, 19900, \ldots$ nth term: $n(2n-1)$

p. 15 DIAGONALS OF A POLYGON

$-, -, 0, 2, 5, 9, 14, 20, 27, \ldots,$
$1175, \ldots,$ nth term: $\frac{1}{2}n(n-3)$

p. 16 CUBIC NUMBERS

$1, 8, 27, 64, 125, 216, 343, \ldots,$
$1728, \ldots 1000000, \ldots$
nth term: n^3

p. 17 POINTS AND LINES

$0, 1, 3, 6, 10, 15, 21, 28, 36, 45,$
$\ldots, 190, \ldots$ nth term: $\frac{1}{2}n(n-1)$

p. 18 SEGMENTS ON A LINE

$0, 1, 3, 6, 10, 15, 21, 28, \ldots,$
$4950, \ldots,$ nth term: $\frac{1}{2}n(n-1)$

p. 19 PROBLEM SUMMARY

1) $1, 4, 9, 25, 36, 49, \ldots, 144,$
   $\ldots, n^2$
2) $1, 3, 6, 10, 15, 21, \ldots, 78,$
   $\ldots, \frac{1}{2}n(n+1)$
3) $2, 6, 12, 20, 30, 42, \ldots, 156,$
   $\ldots, n(n+1)$
4) $1, 5, 12, 22, 35, 51, \ldots, 210,$
   $\ldots, \frac{1}{2}n(3n-1)$
5) $1, 6, 15, 28, 45, 66, \ldots, 276,$
   $\ldots, n(2n-1)$
6) $1, 8, 27, 64, 125, 216, \ldots,$
   $1728, \ldots, n^3$
7) $-, -, 0, 2, 5, 9, 14, 20, \ldots,$
   $\frac{1}{2}n(n-3)$
8) $\frac{1}{2}n(n-1)$ segments
9) $\frac{1}{2}n(n-1)$ segments

SOLUTIONS

Solutions by finite differences.

p.9 WHAT'S MY RULE?

1)

2
$\quad > 2$
4
$\quad > 2$
6
$\quad > 2$
8
$\quad > 2$
10

$a=2, b=0$
$2n$

2)

1
$\quad > 2$
3
$\quad > 2$
5
$\quad > 2$
7
$\quad > 2$
9

$a=2, b=-1$
$2n-1$

3)

3
$\quad > 3$
6
$\quad > 3$
9
$\quad > 3$
12
$\quad > 3$
15

$a=3, b=0$
$3n$

4)

7
$\quad > 3$
10
$\quad > 3$
13
$\quad > 3$
16
$\quad > 3$
19

$a=3, b=4$
$3n+4$

5)

5
$\quad > 5$
10
$\quad > 5$
15
$\quad > 5$
20
$\quad > 5$
25

$a=5, b=0$
$5n$

6)

2
$\quad > 5$
7
$\quad > 5$
12
$\quad > 5$
17
$\quad > 5$
22

$a=5, b=-3$
$5n-3$

7)

$\frac{3}{2}$
$\quad > \frac{1}{2}$
2
$\quad > \frac{1}{2}$
$\frac{5}{2}$
$\quad > \frac{1}{2}$
3
$\quad > \frac{1}{2}$
$\frac{7}{2}$

$a=\frac{1}{2}, b=1$
$\frac{1}{2}n+1$

8)

$-\frac{3}{2}$
$\quad > \frac{1}{2}$
-1
$\quad > \frac{1}{2}$
$-\frac{1}{2}$
$\quad > \frac{1}{2}$
0
$\quad > \frac{1}{2}$
$\frac{1}{2}$

$a=\frac{1}{2}, b=-2$
$\frac{1}{2}n-2$

9)

17
$\quad > 8$
25
$\quad > 8$
33
$\quad > 8$
41
$\quad > 8$
49

$a=8, b=9$
$8n+9$

10)

9
$\quad > 14$
23
$\quad > 14$
37
$\quad > 14$
51
$\quad > 14$
65

$a=14, b=-5$
$14n-5$

p.10 SQUARE NUMBERS

n	T
0	0
1	1
2	4
3	9
4	16
5	25
.	
.	
n	n^2

First differences: >1, >3, >5, >7, >9
Second differences: >2, >2, >2, >2

an^2+bn+c
$a=1, b=0$
$c=0$

p.11 TRIANGULAR NUMBERS

n	T
0	0
1	1
2	3
3	6
4	10
5	15
.	
.	
n	$\frac{1}{2}n^2 + \frac{1}{2}n$ or $\frac{n(n+1)}{2}$

First differences: >1, >2, >3, >4, >5
Second differences: >1, >1, >1, >1

an^2+bn+c
$a=\frac{1}{2}, b=\frac{1}{2}$
$c=0$

p.12 OBLONG NUMBERS

n	O
0	0
1	2
2	6
3	12
4	20
5	30
.	
.	
n	n^2+n or $n(n+1)$

First differences: >2, >4, >6, >8, >10
Second differences: >2, >2, >2, >2

an^2+bn+c
$a=1, b=1$
$c=0$

p.13 PENTAGONAL NUMBERS

n	P
0	0
1	1
2	5
3	12
4	22
5	35
.	
.	
n	$\frac{3}{2}n^2 - \frac{1}{2}n$ or $\frac{n(3n-1)}{2}$

First differences: >1, >4, >7, >10, >13
Second differences: >3, >3, >3, >3

an^2+bn+c
$a=\frac{3}{2}, b=-\frac{1}{2}$
$c=0$

SOLUTIONS

p.14 HEXAGONAL NUMBERS

n	H
0	0
1	1
2	6
3	15
4	28
5	45
.	
.	
.	
n	$2n^2-n$ or $n(2n-1)$

First differences: 1, 5, 9, 13, 17 → second differences: 4, 4, 4, 4

an^2+bn+c
$a=2, b=-1$
$c=0$

p. 15 DIAGONALS OF A POLYGON

n	D
3	0
4	2
5	5
6	9
.	
.	
.	
n	$\frac{1}{2}n^2-\frac{3}{2}n$ or $\frac{n(n-3)}{2}$

First differences: 2, 3, 4 → second differences: 1, 1

an^2+bn+c
$a=\frac{1}{2}, b=-\frac{3}{2}$
$c=0$

p.16 CUBIC NUMBERS

n	C
0	0
1	1
2	8
3	27
4	64
5	125
.	
.	
.	
n	n^3

First differences: 1, 7, 19, 37, 61 → second differences: 6, 12, 18, 24 → third differences: 6, 6, 6

an^3+bn^2+cn+d
$a=1, b=0$
$c=0, d=0$

p.17 POINTS AND LINES

n	S
0	0
1	0
2	1
3	3
4	6
5	10
.	
.	
.	
n	$\frac{1}{2}n^2-\frac{1}{2}n$ or $\frac{n(n-1)}{2}$

First differences: 0, 1, 2, 3, 4 → second differences: 1, 1, 1, 1

an^2+bn+c
$a=\frac{1}{2}, b=-\frac{1}{2}$
$c=0$

p.18 SEGMENTS ON A LINE

Solution identical to Points and Lines solution.

SECTION 2

p.44 WHAT'S MY RULE?

1)

n	t
1	1
2	4
3	9
4	16
5	25
6	36
7	49
8	64
.	
.	
.	
n	n^2

First differences: 3, 5, 7, 9 → second differences: 2, 2, 2

an^2+bn+c
$a=1, b=0$
$c=0$

2)

n	t
1	4
2	7
3	12
4	19
5	28
6	39
7	52
8	67
.	
.	
.	
n	n^2+3

First differences: 3, 5, 7, 9 → second differences: 2, 2, 2

an^2+bn+c
$a=1, b=0$
$c=3$

3)

n	t
1	2
2	6
3	12
4	20
5	30
6	42
7	56
8	72
.	
.	
.	
n	n^2+n

First differences: 4, 6, 8, 10 → second differences: 2, 2, 2

an^2+bn+c
$a=1, b=1$
$c=0$

SOLUTIONS

4)

n	t
1	3
2	12
3	27
4	48
5	75
6	108
7	147
8	192
.	
.	
n	$3n^2$

First differences: 9, 15, 21, 27 Second differences: 6, 6, 6

an^2+bn+c
$a=3, b=0$
$c=0$

5)

n	t
1	−1
2	5
3	15
4	29
5	47
6	69
7	95
8	125
.	
.	
n	$2n^2-3$

First differences: 6, 10, 14, 18 Second differences: 4, 4, 4

an^2+bn+c
$a=2, b=0$
$c=-3$

6)

n	t
1	3
2	7
3	13
4	21
5	31
6	43
7	57
8	73
.	
.	
n	n^2+n+1

First differences: 4, 6, 8, 10 Second differences: 2, 2, 2

an^2+bn+c
$a=1, b=1$
$c=1$

7)

n	t
1	2
2	7
3	16
4	29
5	46
6	67
7	92
8	121
.	
.	
n	$2n^2-n+1$

First differences: 5, 9, 13, 17 Second differences: 4, 4, 4

an^2+bn+c
$a=2, b=-1$
$c=1$

8)

n	t
1	3
2	3
3	5
4	9
5	15
6	23
7	33
8	45
.	
.	
n	n^2-3n+5

First differences: 0, 2, 4, 6 Second differences: 2, 2, 2

an^2+bn+c
$a=1, b=-3$
$c=5$

9)

n	t
1	2
2	5
3	9
4	14
5	20
6	27
7	15
8	44
.	
.	
n	$\frac{1}{2}n^2+\frac{3}{2}n$ or $\frac{1}{2}n(n+3)$

First differences: 3, 4, 5, 6 Second differences: 1, 1, 1

an^2+bn+c
$a=\frac{1}{2}, b=\frac{3}{2}$
$c=0$

SOLUTIONS

10)

n	t
1	0
2	$\frac{4}{3}$
3	4
4	8
5	$\frac{40}{3}$
6	20
7	28
8	$\frac{112}{3}$
.	
.	
n	$\frac{2}{3}n^2 - \frac{2}{3}n$ or $\frac{2n(n-1)}{3}$

First differences: $\frac{4}{3}, \frac{8}{3}, \frac{12}{3}, \frac{16}{3}$; second differences: $\frac{4}{3}, \frac{4}{3}, \frac{4}{3}$
an^2+bn+c, $a=\frac{2}{3}$, $b=-\frac{2}{3}$, $c=0$

p. 45 WHAT'S MY SUM?

1)

n	S
1	1
2	3
3	6
4	10
5	15
.	
.	
n	$\frac{1}{2}n^2 + \frac{1}{2}n$ or $\frac{1}{2}n(n+1)$

First differences: $2, 3, 4, 5$; second differences: $1, 1, 1$
an^2+bn+c, $a=\frac{1}{2}$, $b=\frac{1}{2}$, $c=0$

2)

n	S
1	1
2	4
3	9
4	16
5	25
.	
.	
n	n^2

First differences: $3, 5, 7, 9$; second differences: $2, 2, 2$
an^2+bn+c, $a=1$, $b=0$, $c=0$

3)

n	S
1	3
2	7
3	12
4	18
5	25
.	
.	
n	$\frac{1}{2}n^2 + \frac{5}{2}$ or $\frac{1}{2}n(n+5)$

First differences: $4, 5, 6, 7$; second differences: $1, 1, 1$
an^2+bn+c, $a=\frac{1}{2}$, $b=\frac{5}{2}$, $c=0$

4)

n	S
1	2
2	8
3	18
4	32
5	50
.	
.	
n	$2n^2$

First differences: $6, 10, 14, 18$; second differences: $4, 4, 4$
an^2+bn+c, $a=2$, $b=0$, $c=0$

5)

n	S
1	5
2	13
3	24
4	38
5	55
.	
.	
n	$\frac{3}{2}n^2 + \frac{7}{2}n$ or $\frac{1}{2}n(3n+7)$

First differences: $8, 11, 14, 17$; second differences: $3, 3, 3$
an^2+bn+c, $a=\frac{3}{2}$, $b=\frac{7}{2}$, $c=0$

6)

n	S
1	1
2	6
3	15
4	28
5	45
.	
.	
n	$2n^2-n$ or $n(2n-1)$

First differences: $5, 9, 13, 17$; second differences: $4, 4, 4$
an^2+bn+c, $a=2$, $b=-1$, $c=0$

7)

n	S
1	4
2	13
3	27
4	46
5	70
.	
.	
n	$\frac{5}{2}n^2 + \frac{3}{2}n$ or $\frac{1}{2}n(5n+3)$

First differences: $9, 14, 19, 24$; second differences: $5, 5, 5$
an^2+bn+c, $a=\frac{5}{2}$, $b=\frac{3}{2}$, $c=0$

8)

n	S
1	11
2	32
3	63
4	104
5	155
.	
.	
n	$5n^2+6n$ or $n(5n+6)$

First differences: $21, 31, 41, 51$; second differences: $10, 10, 10$
an^2+bn+c, $a=5$, $b=6$, $c=0$

SOLUTIONS

9)

n	S
1	-5
2	-6
3	-3
4	4
5	15

first differences: $-1,\ 3,\ 7,\ 11$ second differences: $4,\ 4,\ 4$

an^2+bn+c $a=2,\ b=-7$ $c=0$

n	$2n^2-7n$ or $n(2n-7)$

10)

n	S
1	0
2	-2
3	-6
4	-12
5	-20

first differences: $-2,\ -4,\ -6,\ -8$ second differences: $-2,\ -2,\ -2$

an^2+bn+c $a=-1,\ b=1$ $c=0$

n	$-n^2+n$ or $n(1-n)$

11)

n	S
1	$\frac{1}{2}$
2	$\frac{3}{2}$
3	$\frac{6}{2}$
4	$\frac{10}{2}$
5	$\frac{15}{2}$

first differences: $1,\ \frac{3}{2},\ 2,\ \frac{5}{2}$ second differences: $\frac{1}{2},\ \frac{1}{2},\ \frac{1}{2}$

an^2+bn+c $a=\frac{1}{4},\ b=\frac{1}{4}$ $c=0$

n	$\frac{1}{4}n^2+\frac{1}{4}n$ or $\frac{1}{4}n(n+1)$

p.46 CANNONBALL

1) $t_n=\frac{1}{2}n(n+1)$

$t_4=10$

$t_{10}=55$

2) $S_{10}=1+3+6+10+\ldots+55=220$

3)

n	S
1	1
2	4
3	10
4	20
5	35
6	56

first differences: $3,\ 6,\ 10,\ 15,\ 21$ second differences: $3,\ 4,\ 5,\ 6$ third differences: $1,\ 1,\ 1,\ 1$

$a=\frac{1}{6}$ $b=\frac{1}{2}$ $c=\frac{1}{3}$ $d=0$

Therefore, the sum of n triangular numbers, or the nth triangular pyramidal number is

$$\frac{1}{6}n^3+\frac{1}{2}n^2+\frac{1}{3}n = \frac{n(n+1)(n+2)}{6}$$

4) $t_5 = 25,\ t_9 = 81,\ t_n = n^2$

5) $S_9 = 1+4+9+16+\ldots+81 = 285$

6)

n	S
1	1
2	5
3	14
4	30
5	55
6	91

first differences: $4,\ 9,\ 16,\ 25,\ 36$ second differences: $5,\ 7,\ 9,\ 11$ third differences: $2,\ 2,\ 2$

$a=\frac{1}{3}$ $b=\frac{1}{2}$ $c=\frac{1}{6}$ $d=0$

The sum of n square numbers, or the nth square pyramidal number, is

$$\frac{1}{3}n^3+\frac{1}{2}n^2+\frac{1}{6}n = \frac{n(n+1)(2n+1)}{6}$$

p.47 REGIONS IN A CIRCLE

n	R
0	1
1	2
2	4
3	7
4	11
5	16

first differences: $1,\ 2,\ 3,\ 4,\ 5$ second differences: $1,\ 1,\ 1,\ 1$

$a=\frac{1}{2}$ $b=\frac{1}{2}$ $c=1$

n	R
60	1831

n	$\frac{1}{2}n^2+\frac{1}{2}n+1$ or $\frac{n(n+1)}{2}+1$

SOLUTIONS

p.48   TRIANGLES IN A TRIANGLE
(Count only triangles pointing up.)

n	T
0	0
1	1
2	4
3	10
4	20
5	35
.	
.	
n	$\frac{1}{6}n^3+\frac{1}{2}n^2+\frac{1}{3}n$ or $\frac{n(n+1)(n+2)}{6}$

Differences:
0 >1 ; 1 >3 >2 ; 4 >6 >3 >1 ; 10 >10 >4 >1 ; 20 >15 >5 >1

$a=\frac{1}{6}$
$b=\frac{1}{2}$
$c=\frac{1}{3}$
$d=0$

p.49   SQUARES IN A SQUARE

n	S
0	0
1	1
2	5
3	14
4	30
5	55
.	
.	
n	$\frac{1}{3}n^3+\frac{1}{2}n^2+\frac{1}{6}n$ or $\frac{n(n+1)(2n+1)}{6}$

Differences:
0 >1 ; 1 >4 >3 >2 ; 5 >9 >5 >2 ; 14 >16 >7 >2 ; 30 >25 >9 >2

$a=\frac{1}{3}$
$b=\frac{1}{2}$
$c=\frac{1}{6}$
$d=0$

p.50   SQUARES ON A GEOBOARD

(all squares)

n	S
0	0
1	1
2	6
3	20
4	50
5	105
.	
.	
n	$\dfrac{n^4+4n^3+5n^2+2n}{12}$

Differences:
0 >1 >4 ; 1 >5 >9 >5 ; 6 >14 >16 >7 >2 ; 20 >30 >25 >9 >2 ; 50 >55

$a=\frac{1}{12}$
$b=\frac{1}{3}$
$c=\frac{5}{12}$
$d=\frac{1}{6}$
$e=0$

(squares with sides that are not parallel to sides of the geoboard)

n	S
0	0
1	0
2	1
3	6
4	20
5	50
.	
.	
n	$\dfrac{n^4-n^2}{12}$

Differences:
0 >0 >1 ; 0 >1 >4 >3 ; 1 >5 >9 >5 >2 ; 6 >14 >16 >7 >2 ; 20 >30

$a=\frac{1}{12}$
$b=0$
$c=-\frac{1}{12}$
$d=0$
$e=0$

p.51   TEN MEN IN A BOAT
(pairs)

n	M
0	0
1	3
2	8
3	15
4	24
5	35
6	48
.	.
.	.
12	168
.	
n	n^2+2n or $n(n+2)$

Differences:
0 >3 >3 ; 3 >5 >2 ; 8 >7 >2 ; 15 >9 >2 ; 24 >11 >2 ; 35 >13 >2 ; 48

$a=1$
$b=2$
$c=0$

TEN MEN IN A BOAT
(not in pairs)

n	M
0	0
1	1
2	3
3	5
4	8
5	11
6	15

Differences:
0 >1 >1 ; 1 >2 >1 ; 3 >2 >0 ; 5 >3 >1 ; 8 >3 >0 ; 11 >4 >1 ; 15

Note: The method described for "Shifting Pennies" (considering separately odd and even values of n) may also be applied to the solution of "Ten Men in a Boat."

p.52   TOWER OF HANOI

n	M
0	0
1	1
2	3
3	7
4	15
5	31
6	63
7	127

Differences:
0 >1 >1 >1 ; 1 >2 >2 >1 ; 3 >4 >4 >2 ; 7 >8 >8 >4 ; 15 >16 >16 >8 ; 31 >32 >32 >16 ; 63 >64   etc.

The general rule, 2^n-1, is not a polynomial. This problem cannot be solved by finite differences.

The pattern for the number of moves each disc makes is:
$$2^n-1=2^{n-1}+2^{n-2}+2^{n-3}+\ldots+2^0$$

Total   Smallest disc   Largest disc

SOLUTIONS

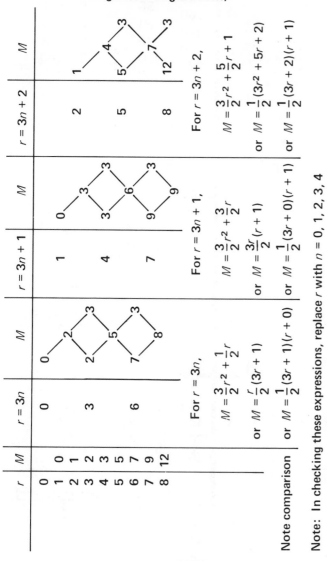

p. 53   SHIFTING PENNIES
(r represents rows added to original configuration.)

For $r = 3n + 2$,
$$M = \frac{3}{2}r^2 + \frac{5}{2}r + 1$$
or $M = \frac{1}{2}(3r^2 + 5r + 2)$ or $M = \frac{1}{2}(3r + 2)(r + 1)$

For $r = 3n + 1$,
$$M = \frac{3}{2}r^2 + \frac{3}{2}r$$
or $M = \frac{3r}{2}(r + 1)$ or $M = \frac{1}{2}(3r + 0)(r + 1)$

For $r = 3n$,
$$M = \frac{3}{2}r^2 + \frac{1}{2}r$$
or $M = \frac{r}{2}(3r + 1)$ or $M = \frac{1}{2}(3r + 1)(r + 0)$

Note comparison

Note: In checking these expressions, replace r with $n = 0, 1, 2, 3, 4$

r	M
0	0
1	1
2	2
3	3
4	5
5	7
6	9
7	12
8	

SECTION 3

p.56   WHAT'S MY RULE?

1)

n	t
1	8
2	27
3	64
4	125
5	216
6	343
7	512
⋮	
n	$n^3 + 3n^2 + 3n + 1$ or $(n+1)^3$

Differences: 19, 37, 61, 91, 127, 169 ; 18, 24, 30, 36, 42 ; 6, 6, 6, 6
$a = 1$ $b = 3$ $c = 3$ $d = 1$

2)

n	t
1	1
2	8
3	27
4	64
5	125
6	216
7	343
⋮	
n	n^3

Differences: 7, 19, 37, 61, 91, 127 ; 12, 18, 24, 30, 36 ; 6, 6, 6, 6
$a = 1$ $b = 0$ $c = 0$ $d = 0$

3)

n	t
1	3
2	11
3	31
4	69
5	131
6	223
7	351
⋮	
n	$n^3 + n + 1$

Differences: 8, 20, 38, 62, 92, 128 ; 12, 18, 24, 30, 36 ; 6, 6, 6, 6
$a = 1$ $b = 0$ $c = 1$ $d = 1$

4)

n	t
1	4
2	19
3	44
4	79
5	124
6	179
7	244
⋮	
n	$5n^2 - 1$

Differences: 15, 25, 35, 45, 55, 65 ; 10, 10, 10, 10, 10
$a = 5$ $b = 0$ $c = -1$

5)

n	t
1	−3
2	11
3	49
4	123
5	245
6	427
7	681
⋮	
n	$2n^3 - 5$

Differences: 14, 38, 74, 122, 182, 254 ; 24, 36, 48, 60, 72 ; 12, 12, 12, 12
$a = 2$ $b = 0$ $c = 0$ $d = -5$

SOLUTIONS

6)

n	t
1	5
2	11
3	19
4	29
5	41
6	55
7	71
⋮	
n	n^2+3n+1

Differences: 6, 8, 10, 12, 14, 16; second differences: 2, 2, 2, 2, 2

$a=1$, $b=3$, $c=1$

7)

n	t
1	−1
2	0
3	1
4	8
5	27
6	64
7	125
⋮	
n	$n^3-6n^2+12n-8$ or $(n-2)^3$

Differences: 1, 1, 7, 19, 37, 61; second: 0, 6, 12, 18, 24; third: 6, 6, 6, 6

$a=1$, $b=-6$, $c=12$, $d=-8$

8)

n	t
1	−2
2	6
3	26
4	64
5	126
6	218
7	346
⋮	
n	n^3+n-4

Differences: 8, 20, 38, 62, 92, 128; second: 12, 18, 24, 30, 36; third: 6, 6, 6, 6

$a=1$, $b=1$, $c=-4$

9)

n	t
1	7
2	25
3	59
4	115
5	199
6	317
7	475
⋮	
n	n^3+2n^2+5n-1

Differences: 18, 34, 56, 84, 118, 158; second: 16, 22, 28, 34, 40; third: 6, 6, 6, 6

$a=1$, $b=2$, $c=5$, $d=-1$

10)

n	t
1	6
2	24
3	60
4	120
5	210
6	336
7	504
⋮	
n	n^3+3n^2+2n or $n(n+1)(n+2)$

Differences: 18, 36, 60, 90, 126, 168; second: 18, 24, 30, 36, 42; third: 6, 6, 6, 6

$a=1$, $b=3$, $c=2$, $d=0$

p.57 WHAT'S MY SUM?

1)

n	S
1	1
2	4
3	10
4	20
5	35
⋮	
n	$\frac{1}{6}n^3+\frac{1}{2}n^2+\frac{1}{3}n$ or $\frac{n(n+1)(n+2)}{6}$

Differences: 3, 6, 10, 15; second: 3, 4, 5; third: 1, 1

$a=\frac{1}{6}$, $b=\frac{1}{2}$, $c=\frac{1}{3}$, $d=0$

2)

n	S
1	1
2	5
3	14
4	30
5	55
⋮	
n	$\frac{1}{3}n^3+\frac{1}{2}n^2+\frac{1}{6}n$ or $\frac{n(n+1)(2n+1)}{6}$

Differences: 4, 9, 16, 25; second: 5, 7, 9; third: 2, 2

$a=\frac{1}{3}$, $b=\frac{1}{2}$, $c=\frac{1}{6}$, $d=0$

3)

n	S
1	1
2	6
3	18
4	40
5	75
⋮	
n	$\frac{1}{2}n^3+\frac{1}{2}n^2$ or $\frac{n^2(n+1)}{2}$

Differences: 5, 12, 22, 35; second: 7, 10, 13; third: 3, 3

$a=\frac{1}{2}$, $b=\frac{1}{2}$, $c=0$, $d=0$

SOLUTIONS

4)

n	S			
1	1			
2	7	6		
3	22	15	9	
4	50	28	13	4
5	95	45	17	4
.				
.				

$a=\frac{2}{3}$ $b=\frac{1}{2}$ $c=-\frac{1}{6}$ $d=0$

$$n \quad \frac{2}{3}n^3+\frac{1}{2}n^2-\frac{1}{6}n \quad \text{or} \quad \frac{n(n+1)(4n-1)}{6}$$

5)

n	S			
1	3			
2	10	7		
3	23	13	6	
4	44	21	8	2
5	75	31	10	2
.				
.				

$a=\frac{1}{3}$ $b=1$ $c=\frac{5}{3}$ $d=0$

$$n \quad \frac{1}{3}n^3+n^2+\frac{5}{3}n \quad \text{or} \quad \frac{n(n^2+3n+5)}{3}$$

6)

n	S			
1	2			
2	8	6		
3	20	12	6	
4	40	20	8	2
5	70	30	10	2
.				
.				

$a=\frac{1}{3}$ $b=1$ $c=\frac{2}{3}$ $d=0$

$$n \quad \frac{1}{3}n^3+n^2+\frac{2}{3}n \quad \text{or} \quad \frac{n(n+1)(n+2)}{3}$$

7)

n	S			
1	13			
2	42	29		
3	93	51	22	
4	172	79	28	6
5	285	113	34	6
.				
.				

$a=1$ $b=5$ $c=7$ $d=0$

$$n \quad n^3+5n^2+7n$$

8)

n	S			
1	$\frac{3}{2}$			
2	$\frac{11}{2}$	4		
3	13	$\frac{15}{2}$	$\frac{7}{2}$	
4	25	12	$\frac{9}{2}$	1
5	$\frac{85}{2}$	$\frac{35}{2}$	$\frac{11}{2}$	1
.				
.				

$a=\frac{1}{6}$ $b=\frac{3}{4}$ $c=\frac{7}{12}$ $d=0$

$$n \quad \frac{1}{6}n^3+\frac{3}{4}n^2+\frac{7}{12}n \quad \text{or} \quad \frac{n(n+1)(2n+7)}{12}$$

9)

n	S				
1	1				
2	9	8			
3	36	27	19		
4	100	64	37	18	
5	225	125	61	24	6
6	441	216	91	30	6
.					
.					

$a=\frac{1}{4}$ $b=\frac{1}{2}$ $c=\frac{1}{4}$ $d=0$ $e=0$

$$n \quad \frac{1}{4}n^4+\frac{1}{2}n^3+\frac{1}{4}n^2 \quad \text{or} \quad \left[\frac{1}{2}n(n+1)\right]^2$$

10)

n	S				
1	1				
2	28	27			
3	153	125	98		
4	496	343	218	120	
5	1225	729	386	168	48
6	2556	1331	602	216	48
.					
.					

$a=2$ $b=0$ $c=-1$ $d=0$ $e=0$

$$n \quad 2n^4-n^2 \quad \text{or} \quad n^2(2n^2-1)$$

p.58   WHAT'S MY SUM?

1)

n	S				
1	1				
2	10	9			
3	35	25	16		
4	84	49	24	8	
5	165	81	32	8	
6	334	121			
.					
.					

$a=\frac{4}{3}$ $b=0$ $c=-\frac{1}{3}$ $d=0$

$$n \quad \frac{4}{3}n^3-\frac{1}{3}n \quad \text{or} \quad \frac{n(2n+1)(2n-1)}{3}$$

95

SOLUTIONS

2)

n	S
1	4
2	20
3	56
4	120
5	220
6	364

16 > 20
36 > 28 > 8
64 > 36 > 8
100 > 44 > 8
144

$a=\frac{4}{3}$
$b=2$
$c=\frac{2}{3}$
$d=0$

$n \quad \frac{4}{3}n^3+2n^2+\frac{2}{3}n \ \text{ or } \ \frac{2n(2n+1)(n+1)}{3}$

3)

n	S
1	1
2	17
3	98
4	354
5	979
6	2275

16 > 65
81 > 175 > 110
256 > 369 > 194 > 84
625 > 671 > 302 > 108 > 24
1296

$n \quad \frac{1}{5}n^5+\frac{1}{2}n^4+\frac{1}{3}n^3-\frac{1}{30}n$

$a=\frac{1}{5}$
$b=\frac{1}{2}$
$c=\frac{1}{3}$
$d=0$
$e=\frac{1}{30}$
$f=0$

4)

n	S
1	3
2	11
3	26
4	50
5	85
6	133

8 > 7
15 > 9 > 2
24 > 11 > 2
35 > 13 > 2
48

$a=\frac{1}{3}$
$b=\frac{3}{2}$
$c=\frac{7}{6}$
$d=0$

$n \quad \frac{1}{3}n^3+\frac{3}{2}n^2+\frac{7}{6}n \ \text{ or } \ \frac{n(n+1)(2n+7)}{6}$

5)

n	S
1	6
2	26
3	68
4	140
5	250
6	406

20 > 22
42 > 30 > 8
72 > 38 > 8
110 > 46 > 8
156

$a=\frac{4}{3}$
$b=3$
$c=\frac{5}{3}$
$d=0$

$n \quad \frac{4}{3}n^3+3n^2+\frac{5}{3}n \ \text{ or } \ \frac{n(n+1)(4n+5)}{3}$

6)

n	S
1	$\frac{1}{3}$
2	$\frac{4}{3}$
3	$\frac{9}{3}$
4	$\frac{16}{3}$
5	$\frac{25}{3}$
6	$\frac{36}{3}$

$\frac{3}{3} > \frac{2}{3}$
$\frac{5}{3} > \frac{2}{3}$
$\frac{7}{3} > \frac{2}{3}$
$\frac{9}{3} > \frac{2}{3}$
$\frac{11}{3}$

$a=\frac{1}{3}$
$b=0$
$c=0$

$n \quad \frac{1}{3}n^2$

7)

n	S
1	6
2	30
3	90
4	210
5	420
6	756

24 > 36
60 > 60 > 24
120 > 90 > 30 > 6
210 > 126 > 36 > 6
336

$n \quad \frac{1}{4}n^4+\frac{3}{2}n^3+\frac{11}{4}n^2+\frac{3}{2}n \ \text{ or }$

$\frac{n(n+1)(n+2)(n+3)}{4}$

$a=\frac{1}{4}$
$b=\frac{3}{2}$
$c=\frac{11}{4}$
$d=\frac{3}{2}$
$e=0$

8)

n	S
1	1
2	4
3	9
4	16
5	25
6	36

3 > 2
5 > 2
7 > 2
9 > 2
11

$a=1$
$b=0$
$c=0$

$n \quad n^2$

SOLUTIONS

9)

n	S
1	$\frac{1}{2}$
2	$\frac{2}{3}$
3	$\frac{3}{4}$
4	$\frac{4}{5}$
5	$\frac{5}{6}$
6	$\frac{6}{7}$
.	
.	
.	
n	$\frac{n}{n+1}$

This problem cannot be solved by finite differences because the general rule is not a polynomial. However, the pattern is obvious.

10)

n	S
1	$\frac{1}{3}$
2	$\frac{2}{5}$
3	$\frac{3}{7}$
4	$\frac{4}{9}$
5	$\frac{5}{11}$
6	$\frac{6}{13}$
.	
.	
.	
n	$\frac{n}{2n+1}$

This problem cannot be solved by finite differences because the general rule is not a polynomial. However, the pattern is obvious.

p.59   RECTANGLES IN A SQUARE

n	R
0	0
1	1
2	9
3	36
4	100
5	225
6	441
7	784
8	1296
.	
.	
25	90,000
.	
.	
n	

$1 > 1$
$9 > 8 > 7$
$36 > 27 > 19 > 12$
$100 > 64 > 37 > 18 > 6$
$225 > 125 > 61 > 24 > 6$
$441 > 216 > 91 > 30 > 6$
$784 > 343 > 127 > 36 > 6$
$1296 > 512 > 169 > 42 > 6$

$an^4 + bn^3 + cn^2 + dn + e$

$a = \frac{1}{4}, b = \frac{1}{2}, c = \frac{1}{4}, d = 0, e = 0$

$\frac{1}{4}n^4 + \frac{1}{2}n^3 + \frac{1}{4}n^2 = \frac{n(n+1)}{2} \cdot \frac{n(n+1)}{2}$

$$= \frac{n^2}{4}(n+1)^2$$

This polynomial is the same as the sum of the cubes of the first n positive integers. See problem 9 on page 57.

p. 60   RED-FACED CUBES

e	no. of unit cubes	three red faces	two red faces
2	8	8	0
3	27	8	12
4	64	8	24
5	125	8	36
6	216	8	48
7	343	8	60
.			
.			
e	e^3	8	$12(e-2)$ (See below)

e	one red face	no red faces
2	0	0
3	6	1
4	24	8
5	54	27
6	96	64
7	150	125
.		
.		
e	$6(e-2)^2$ (See below)	$(e-2)^3$

two red faces

$0 > 12$
$12 > 12$
$24 > 12$
$36 > 12$
$48 > 12$
60

one red face

$0 > 6$
$6 > 18 > 12$
$24 > 30 > 12$
$54 > 42 > 12$
$96 > 54 > 12$
150

two red faces	one red face
a=12	a=6
b=−24	b=−24
12e−24	c=24
or 12(e−2)	$6e^2 − 24e + 24$
	or $6(e−2)^2$

Check: Show that
$$e^3 = 8 + 12(e-2) + 6(e-2)^2 + (e-2)^3$$
is true.

SOLUTIONS

p.61 MOSAICS

1) Area of square = 9

2) Area of hexagon = $6\sqrt{3} \approx 10.392$

3) Hexagon. Hexagonal cell.

4) Six sides. Note: The plane cannot be filled completely using only regular octagons, decagons, or dodecagons, and space cannot be filled using the related prisms.

5) MOSAIC OF HEXAGONS

n	S
1	1
2	7
3	19
4	37
5	61

> 6 > 6
> 12 > 6
> 18 > 6
> 24

$a=3$
$b=-3$
$c=1$
$S=3n^2-3n+1$

6) No. If $n=6$, $S=91 = 7\cdot13$

7) MOSAIC OF SQUARES

n	S
1	1
2	5
3	13
4	25
5	41

> 4 > 4
> 8 > 4
> 12 > 4
> 16

$a=2$
$b=-2$
$c=1$
$S=2n^2-2n+1$

8) Yes. When $n=21$, $S=841=29^2$

9) MOSAIC OF TRIANGLES

n	S
1	1
2	4
3	10
4	19
5	31

> 3 > 3
> 6 > 3
> 9 > 3
> 12

$a=\frac{3}{2}$
$b=-\frac{3}{2}$
$c=1$
$S=\frac{1}{2}(3n^2-3n+2)$

10) Yes. When $n=10$, $S=136$, which is the 16th triangular number.

11) DESIGN ON BACK COVER
Solution for the sequence:

n	t
1	12
2	30
3	48
4	66
5	84

> 18
> 18
> 18
> 18

$a=18$
$b=-6$
$t=18n-6$ or $6(3n-1)$

Solution for the series:

n	S
1	12
2	42
3	90
4	156
5	240

> 30 > 18
> 48 > 18
> 66 > 18
> 84 > 18

$a=9$
$b=3$
$c=0$

Therefore, $S=9n^2+3n$
$=3n(3n+1)$

p.64 TRIANGLE FENCES AND CORRALS

1)

S	F
7	15
9	21
11	27
13	33
15	39

$F=3S-6$

2)

S	I
7	1
9	4
11	9
13	16
15	25

> 3 > 2
> 5 > 2
> 7 > 2
> 9

$49a+7b+c=1$
$81a+9b+c=4$
$121a+11b+c=9$

$a=\frac{1}{4}$, $b=\frac{-5}{2}$, $c=\frac{25}{4}$
$I=\frac{1}{4}(S-5)^2$

3)

S	$F+I$
7	16
9	25
11	36
13	49
15	64

> 9 > 2
> 11 > 2
> 13 > 2
> 15

$49a+7b+c=16$
$81a+9b+c=25$
$121a+11b+c=36$

$a=\frac{1}{4}$, $b=\frac{1}{2}$, $c=\frac{1}{4}$
$F+I=\frac{1}{4}(S+1)^2$

Check: Show that
$\frac{1}{4}(S-5)^2+(3S-6)=\frac{1}{4}(S+1)^2$ is true

SOLUTIONS

4) Yes.
   Solving $\frac{1}{4}(S-5)^2 > 3S-6$
   gives $S > 11+6\sqrt{2}$,
   or $S > 19.5$ approximately.
   Therefore, $I > F$ for $S > 21$.

5) No.

6)

	No. of triangles in each row			No. of triangles on or above each row		
Row	Red	White	Total	Red	White	Total
1	1		1	1		1
2	2	1	3	3	1	4
3	3	2	5	6	3	9
4	4	3	7	10	6	16
5	5	4	9	15	10	25
6	6	5	11	21	15	36

7) The sum of two consecutive <u>triangle</u> numbers is a square number. i.e. triangle no.$_2$ + triangle no.$_1$ = square no.$_2$.

8) Yes. When $S=17$, $F=45$, and $I=36$.

p.67 POLYGONAL NUMBERS

	t_1	t_2	t_3	t_4	t_5	
Triangular $S=3$	1,	3,	6,	10,	15, ...	$t_n : \frac{n^2+n}{2}$
Square $S=4$	1,	4,	9,	16,	25, ...	$t_n : \frac{2n^2+0n}{2}$
Pentagonal $S=5$	1,	5,	12,	22,	35, ...	$t_n : \frac{3n^2-n}{2}$
Hexagonal $S=6$	1,	6,	15,	28,	45, ...	$t_n : \frac{4n^2-2n}{2}$
Heptagonal $S=7$	1,	7,	18,	34,	55, ...	$t_n : \frac{5n^2-3n}{2}$
Octagonal $S=8$	1,	8,	21,	40,	65, ...	$t_n : \frac{6n^2-4n}{2}$

	t_1	t_2	t_3	t_4	t_5	
Nonagonal $S=9$	1,	9,	24,	46,	75, ...	$t_n : \frac{7n^2-5n}{2}$
Decagonal $S=10$	1,	10,	27,	52,	85, ...	$t_n : \frac{8n^2-6n}{2}$
$\vdots$						
S	1,	S,	$3(S-1)$,	$2(3S-4)$,	$5(2S-3)$, ...	$t_n : \frac{(S-2)n^2-(S-4)n}{2}$

p.68 COMPARING SERIES

1) Yes
2) $\bigcirc_{n+1}$
3) The sum of $\square$ Mosaic numbers = $(n^2-2n+1) + n^2 = 2n^2-2n+1$

4)

n	t_n				
1	0				
2	3	> 3			$a=\frac{2}{3}$
3	13	> 10	> 7		$b=-\frac{1}{2}$
4	34	> 21	> 11	> 4	$c=-\frac{1}{6}$
5	70	> 36	> 15	> 4	$d=0$
6	125	> 55	> 19	> 4	

$t_n = \frac{2}{3}n^3 - \frac{1}{2}n^2 - \frac{1}{6}n = \frac{n(n-1)(4n+1)}{6}$

Check: Show that
$n^3 - \frac{n(n+1)(2n+1)}{6} = \frac{n(n-1)(4n+1)}{6}$
is true.

p.70 SERIES SYNTHESIS

1) $t_n = 1+(n-1)3 = 3n-2$
2) $S_n = \frac{n}{2}(1+3n-2) = \frac{n}{2}(3n-1)$
3) Pentagonal numbers
4) $t_n = \frac{n}{2}(3n-1)$ Note: This is the same as the sum in step (3).

5)

n	S				
1	1				$a=\frac{1}{2}$
2	6	> 5			$b=\frac{1}{2}$
3	18	> 12	> 7		$c=0$
4	40	> 22	> 10	> 3	$d=0$
5	75	> 35	> 13	> 3	

$S = \frac{n}{2}^3 + \frac{n}{2}^2(n+1)$

SOLUTIONS

6) $t_n = \frac{n^2}{2}(n+1)$. This is the same as the sum in step 5.

p. 72 PYRAMIDAL NUMBERS

	t_1	t_2	t_3	t_4	t_5

$S=3$ 1, 4, 10, 20, 35, . . .
$$t_n : \frac{n(n+1)(n+2)}{6}$$

$S=4$ 1, 5, 14, 30, 55, . . .
$$t_n : \frac{n(n+1)(2n+1)}{6}$$

$S=5$ 1, 6, 18, 40, 75, . . .
$$t_n : \frac{n(n+1)(3n+0)}{6}$$

$S=6$ 1, 7, 22, 50, 95, . . .
$$t_n : \frac{n(n+1)(4n-1)}{6}$$

$S=7$ 1, 8, 26, 60, 115, . . .
$$t_n : \frac{n(n+1)(5n-2)}{6}$$

$S=8$ 1, 9, 30, 70, 135, . . .
$$t_n : \frac{n(n+1)(6n-3)}{6}$$

$S=9$ 1, 10, 34, 80, 155, . . .
$$t_n : \frac{n(n+1)(7n-4)}{6}$$

$S=10$ 1, 11, 38, 90, 175, . . .
$$t_n : \frac{n(n+1)(8n-5)}{6}$$

.
.
.

S 1, $(S+1)$, $2(2S-1)$, $10(S-1)$, $5(4S-5)$, . . .
$$t_n : \frac{n(n+1)[(S-2)n-(S-5)]}{6}$$

p. 73 PYTHAGOREAN TRIPLES

1) **Column a**
  21st term: 41
  nth term: $2n-1$

  Column b
  21st term: 840
  nth term: $2n(n-1)$

  0
  $\underset{4}{\overset{}{>}}$ 4
  4 $\underset{8}{\overset{}{>}}$ $>$ 4
  12 $\underset{12}{\overset{}{>}}$ $>$ 4
  24 $\underset{16}{\overset{}{>}}$ $>$ 4
  40

  $a=2, b=-2,$
  $c=0$

Column c
21st term: 841
nth term: $2n^2-2n+1$

Check: Show that
$(2n-1)^2+(2n^2-2n)^2 = (2n^2-2n+1)^2$ is true.
$n=21 \rightarrow (41 - 840 - 841)$

2) **Column a**
  10th term: 40
  nth term: $4n$

  Column b
  10th term: 399
  nth term: $4n^2-1 = (2n+1)(2n-1)$

  3
  $\underset{15}{\overset{12}{>}}$
  15 $\underset{20}{\overset{}{>}}$ $>$ 8
  35 $\underset{28}{\overset{}{>}}$ $>$ 8
  63 $\underset{36}{\overset{}{>}}$ $>$ 8
  99

  $a=4, b=0$
  $c=-1$

  Column c
  10th term: 401
  nth term: $(4n^2-1)+2=4n^2+1$

  Check: Show that
  $(4n)^2+(4n^2-1)^2=(4n^2+1)^2$ is true.
  $n=10 \rightarrow (40-399-401)$

3) **Column a**
  12th term: 100
  nth term: $12+(n-1)\,8=8n+4$
       $=4(2n+1)$

  Column b
  12th term: 621
  nth term: $4n^2+4n-3$
       $=(2n+3)(2n-1)$

  5
  $\underset{21}{\overset{16}{>}}$
  21 $\underset{24}{\overset{}{>}}$ $>$ 8
  45 $\underset{32}{\overset{}{>}}$ $>$ 8
  77 $\underset{40}{\overset{}{>}}$ $>$ 8
  117

  $a=4, b=4,$
  $c=-3$

SOLUTIONS

Column c

12th term: 629

nth term: $(4n^2+4n-3)+8$
$$= 4n^2+4n+5$$

Check: Show that
$$(8n+4)^2+(4n^2+4n-3)^2 =$$
$$(4n^2+4n+5)^2 \text{ is true.}$$
$$n=12 \rightarrow (100-621-629)$$

4) Column a

nth term: $15+(n-1)6=$
$$6n+9=3(2n+3)$$

Column b

nth term: $2n^2+6n=2n(n+3)$

$$
\begin{array}{l}
8 \\
20 \\
36 \\
56 \\
80
\end{array}
\begin{array}{l}
> 12 \\
> 16 \\
> 20 \\
> 24
\end{array}
\begin{array}{l}
> 4 \\
> 4 \\
> 4
\end{array}
\quad
\begin{array}{l}
a=2 \\
b=6 \\
c=0
\end{array}
$$

Column c

nth term: $2n^2+6n+9$

Check: Show that
$$(6n+9)^2+(2n^2+6n)^2 =$$
$$(2n^2+6n+9)^2 \text{ is true.}$$

These are not primitive:

$n=3 \rightarrow 9(3-4-5)$
$n=6 \rightarrow 9(5-12-13)$
$n=9 \rightarrow 9(7-24-25)$

p.74 PICK'S THEOREM

3 boundary points

N_i	A	
0	$\frac{1}{2}$	> 1
1	$\frac{3}{2}$	> 1
2	$\frac{5}{2}$	> 1
3	$\frac{7}{2}$	> 1
4	$\frac{9}{2}$	
·		
·		
N_i	$N_i + \frac{1}{2}$	

$a=1$
$b=\frac{1}{2}$

4 boundary points

N_i	A	
0	1	> 1
1	2	> 1
2	3	> 1
3	4	> 1
4	5	
·		
·		
N_i	$N_i + 1$	

$a=1$
$b=1$

5 boundary points

N_i	A	
0	$\frac{3}{2}$	> 1
1	$\frac{5}{2}$	> 1
2	$\frac{7}{2}$	> 1
3	$\frac{9}{2}$	> 1
4	$\frac{11}{2}$	
·		
·		
N_i	$N_i + \frac{3}{2}$	

$a=1$
$b=\frac{3}{2}$

N_b boundary points

N_b	A	
3	$i + \frac{1}{2}$	$> \frac{1}{2}$
4	$i + 1$	$> \frac{1}{2}$
5	$i + \frac{3}{2}$	$> \frac{1}{2}$
6	$i + 2$	$> \frac{1}{2}$
·		
·		
N_b	$\frac{1}{2}N_b + i - 1$	

$a=\frac{1}{2}$
$b=i-1$

p.76 GRASSHOPPER GAMES

1) Note: The minimum number of moves uses the diagonal jump as much as possible.

$p=1$

S	M	
2	1	> 1
3	2	> 1
4	3	> 1
5	4	

$a=1, b=-1$
$M=S-1$ for
$S \geqslant 2$

2) $p=4$

S	M	
3	3	> 2
4	5	> 2
5	7	> 2
6	9	> 2
7	11	> 2
8	13	

$a=2, b=-3$
$M=2S-3$ for
$S \geqslant 3$

3) $p=9$

S	M	
5	12	> 5
6	17	> 5
7	22	> 5
8	27	> 5
9	32	

$a=5, b=-13$
$M=5S-13$ for
$S \geqslant 5$

SOLUTIONS

4) $p=16$

S	M
6	20
7	28
8	36
9	44

> 8
> 8
> 8

$a=8, b=-28$
$M=8S-28$ for
$S \geqslant 6$

5) For $p=9$ and 16, patterns are established if $S-\sqrt{p} \geqslant 2$.

p.78 CIRCLES

1)

P	C
1	0
2	1
3	3
4	6
5	10
6	15

$> 1 > 1$
$> 2 > 1$
$> 3 > 1$
$> 4 > 1$
> 5

$a=\frac{1}{2}, b=-\frac{1}{2}, c=0$
$C=\frac{1}{2}P^2 -\frac{1}{2}P$
$=\frac{P}{2}(P-1)$

2) Solution for problem 1 is the same as for segments formed by n non-collinear points. $S=\frac{1}{2}n(n-1)$. See page 17.

3)

C	R
0	1
1	2
2	4
3	7
4	11
5	16
6	22

$> 1 > 1$
$> 2 > 1$
$> 3 > 1$
$> 4 > 1$
$> 5 > 1$
> 6

$a=\frac{1}{2}, b=\frac{1}{2}, c=1$
$R=\frac{1}{2}C^2 +\frac{1}{2}C+1$
$=\frac{1}{2}(C^2 +C+2)$

4) a. Parallel chords: $R=C+1$
   b. Chords with common end point: $R=C+1$
   c. Chords concurrent in interior of circle: $R=2C$
   d. Each chord intersecting two other chords: $R=2C$

5)

C	I
1	0
2	1
3	3
4	6
5	10
6	15

$I=\frac{1}{2}C^2 -\frac{1}{2}C$
$=\frac{1}{2}C(C-1)$

6) See solution for problem 1.

7)

C	S
1	1
2	4
3	9
4	16
5	25
6	36

$S=C^2$

8) $I=\frac{1}{2}C^2 -\frac{1}{2}C$ and $S=C^2$
   $I=\frac{1}{2}S-\frac{1}{2}C$ or $2I=S-C$
   Therefore, $S=C+2I$

   This relation is valid for C chords, no three of which are concurrent, and which intersect at I interior points.

9)

P	I
1	0
2	0
3	0
4	1
5	5
6	15
7	35
8	70

$> 0 > 0$
$> 0 > 0 > 1 > 1$
$> 0 > 1 > 3 > 2 > 1$
$> 1 > 4 > 6 > 3 > 1$
$> 5 > 10 > 10 > 4 > 1$
$> 15 > 20 > 15 > 5$
$> 35 > 35$

$I=\frac{P}{24}(P^3 -6P^2 +11P-6)$
$=\frac{P}{24}(P-1)(P-2)(P-3)$

$a=\frac{1}{24}$
$b=\frac{-1}{4}$
$c=\frac{11}{24}$
$d=\frac{-1}{4}$
$e=0$

Alternate solution: Any four points on the circle determine a quadrilateral whose diagonals intersect at one point in the interior. For each set of four of the P points, there is

102

SOLUTIONS

one and only one such intersection. Therefore,

$$I = \binom{P}{4} = \frac{P(P-1)(P-2)(P-3)}{4!}$$

10)

Chords from point	Number of intersections	
	if P=7	if P=8
A	0	
B	1+2+3+4	+5
C	2+4+6	+8
D	3+6	+9
E	4	+8
F	0	+5
G	0	
H		0

11)

P	R					
0	1					
1	1	0				
2	2	1	1	0		
3	4	2	1	1	0	
4	8	4	2	1	1	1
				2	1	
5	16	8	4	3	1	
		15	7		1	
6	31		11	4		
7	57	26				

$a=\frac{1}{24}$

$b=\frac{-1}{4}$

$c=\frac{23}{24}$

$d=\frac{-3}{4}$

$e=1$

$$R=\frac{1}{24}(P^4-6P^3+23P^2-18P+24)$$

This solution is equivalent to

$$R=\binom{P-1}{0}+\binom{P-1}{1}+\binom{P-1}{2}+\binom{P-1}{3}+\binom{P-1}{4}$$

This is an excellent example of a situation where insufficient data may lead to a false conclusion. Students might assume that $R=32$ when $P=6$.

See R. A. Gibbs in bibliography.

12) INTERSECTING CIRCLES

n	R		
1	1		
2	3	2	
3	7	4	2
4	13	6	2
5	21	8	2
6	31	10	2

$a=1$, $b=-1$, $c=1$

$R=n^2-n+1$

Note: Given that the circles are congruent and the arrangement symmetrical, segments joining points of intersection of the circles form two triangles if $n=3$, three squares if $n=4$, four pentagons if $n=5$, and five hexagons if $n=6$.

13) Including the exterior of the circles, $R' = n^2-n+2$

p.82 NETWORKS

1) $R + P - L = 1$

Note: Below, n represents the number of points on a side of the polygon.

2) Triangles

n	R	P	L
1	0	1	0
2	1	3	3
3	4	6	9
4	9	10	18
5	16	15	30
6	25	21	45

$R=(n-1)^2$

$P=\frac{1}{2}n(n+1)$, the triangular numbers.

L			
0			
3	3	3	
9	6	3	
18	9	3	
30	12		

$a=\frac{3}{2}$

$b=\frac{-3}{2}$

$c=0$

$L=\frac{3}{2}n(n-1)$

Check: Show that
$(n-1)^2+\frac{1}{2}n(n+1)-\frac{3}{2}n(n-1)=1$ is true.

3) Squares

n	R	P	L
1	0	1	0
2	1	4	4
3	4	9	12
4	9	16	24
5	16	25	40
6	25	36	60

SOLUTIONS

$R=(n-1)^2$

$P=n^2$

$$
\begin{array}{c}
\underline{L} \\
0 \\
4 \\
12 \\
24 \\
40
\end{array}
\qquad
\begin{array}{c}
4 \\
8 \\
12 \\
16
\end{array}
\qquad
\begin{array}{c}
4 \\
4 \\
4
\end{array}
$$

$a=2$

$b=-2$

$c=0$

$L=2n(n-1)$

Check: Show that

$(n-1)^2 + n^2 - 2n(n-1) = 1$ is true.

4) Pentagons

n	R	P	L
1	0	1	0
2	1	5	5
3	2	12	13
4	3	22	24
5	4	35	38
6	5	51	55

$R=n-1$

$P=\frac{1}{2}n(3n-1)$, the pentagonal numbers.

$$
\begin{array}{c}
\underline{L} \\
0 \\
5 \\
13 \\
24 \\
38
\end{array}
\qquad
\begin{array}{c}
5 \\
8 \\
11 \\
14
\end{array}
\qquad
\begin{array}{c}
3 \\
3 \\
3
\end{array}
$$

$a=\frac{3}{2}$

$b=\frac{1}{2}$

$c=-2$

$L=\frac{1}{2}(n-1)(3n+4)$

Check: Show that

$(n-1)+\frac{1}{2}n(3n-1)-\frac{1}{2}(n-1)(3n+4)=1$

is true.

5) Hexagons

n	R	P	L
1	0	1	0
2	1	6	6
3	2	15	16
4	3	28	30
5	4	45	48
6	5	66	70

$R=n-1$

$P=n(2n-1)$, the hexagonal numbers.

$$
\begin{array}{c}
\underline{L} \\
0 \\
6 \\
16 \\
30 \\
48
\end{array}
\qquad
\begin{array}{c}
6 \\
10 \\
14 \\
18
\end{array}
\qquad
\begin{array}{c}
4 \\
4 \\
4
\end{array}
$$

$a=2$

$b=0$

$c=-2$

$L=2(n-1)(n+1)$

Check: Show that

$(n-1)+n(2n-1)-2(n-1)(n+1)=1$

is true.

6) Yes, using $R=0$

7) Intersecting Circles

n	P
1	0
2	2
3	6
4	12
5	20
6	30

$$
\begin{array}{c}
2 \\
4 \\
6 \\
8 \\
10
\end{array}
\qquad
\begin{array}{c}
2 \\
2 \\
2 \\
2
\end{array}
$$

$a=1$

$b=-1$

$c=0$

$P=n(n-1)$

$L=2n(n-1)$

This is the same relation between n and L found in problem 3 for squares.

Check: Show that $R+P-L=1$

i.e. $(n^2-n+1)+(n^2-n)-(2n^2-2n)=1$.

p.84 MULTIPATHS

1) $w=1$

ℓ	P
1	2
2	3
3	4
4	5
5	6

$P=\ell+1$

SOLUTIONS

2) $w=2$

ℓ	P
1	3
2	6
3	10
4	15
5	21

$3 > 1$
$4 > 1$
$5 > 1$
$6 > 1$

$a=\frac{1}{2}$, $b=\frac{3}{2}$, $c=1$

$P=\frac{1}{2}(\ell+1)(\ell+2)$

3) $w=3$

ℓ	P
1	4
2	10
3	20
4	35
5	56

$6 > 4 > 1$
$10 > 5 > 1$
$15 > 6 > 1$
21

$a=\frac{1}{6}$

$b=1$

$c=\frac{11}{6}$

$d=1$

$P=\frac{1}{6}(\ell+1)(\ell+2)(\ell+3)$

Note: There is an alternate solution. For $w=3$ and $\ell=5$, the total trip is 8 units. Therefore,

$P=\frac{8!}{3!\cdot5!} = \frac{8\cdot7\cdot6}{3\cdot2} = 56$, etc.

4) The grid is Pascal's triangle (a right triangle).

1	7	28	84	210		
1	6	21	56	126	252	
1	5	15	35	70	126	210
1	4	10	20	35	56	84
1	3	6	10	15	21	28
1	2	3	4	5	6	7
1	1	1	1	1	1	

5) $w=4$

(See fourth column of grid in solution for problem 4.)

ℓ	P
1	5
2	15
3	35
4	70
5	126
6	210

$10 > 10 > 5 > 1$
$20 > 15 > 6 > 1$
$35 > 21 > 7 > 1$
$56 > 28$
84

$a=\frac{1}{24}$

$b=\frac{5}{12}$

$c=\frac{35}{24}$

$d=\frac{25}{12}$

$e=1$

$P=\frac{1}{24}(\ell+1)(\ell+2)(\ell+3)(\ell+4)$

SUPPLEMENTARY MATERIAL

"The whole of mathematics consists in the organization of a series of aids to the imagination in the process of reasoning."
A. N. Whitehead

■ **Glossary**

■ **Reference Charts**

■ **Bibliography**

GLOSSARY

Arithmetic Sequence — An arithmetic sequence is a set of numbers in which each term, after the first, is formed by adding a fixed number to the preceding term. The fixed number is called the common difference.

Chord — A chord of a circle is a segment whose end points lie on the circle.

Collinear — A set of points is collinear if a straight line may be drawn through all of them.

Concurrent — Two or more lines are concurrent if they have a common point.

Conjecture — To make a guess, or to make an inference from incomplete evidence, is to make a conjecture.

Counting Numbers (or natural numbers) — The set of counting numbers is an arithmetic sequence in which the first term is one and the common difference is one.

Cubic Numbers — Cubic numbers are the set of cubes of the natural (counting) numbers.

Dodecahedron — A dodecahedron is a polyhedron with twelve faces that are either pentagons or rhombuses. The latter is called a rhombic dodecahedron.

Equilateral Triangle — An equilateral triangle is a triangle whose sides have the same length.

Figurate Numbers — The set of figurate numbers contains linear numbers, polygonal numbers, pyramidal numbers, and polyhedral numbers. See page 111.

Finite (sequence) — A sequence is finite if it has a last term.

Finite Differences — Finite Differences is a problem solving technique used to find the general (nth) term of a patterned sequence of numbers.

Generalization — A generalization is the rule which applies to all cases of a certain type of problem. When a generalization is expressed in equation form, it is the nth, or *general*, term.

Geoboard — A geoboard is a physical model of an array of points. Points, usually represented by plastic or metal pegs, are attached to a base.

Geometric Sequence — A geometric sequence is a set of numbers in which each term, after the first, is formed by multiplying the preceding term by a fixed number, called the common ratio.

GLOSSARY

Hexagonal Number — The number of objects that can be arranged in a hexagonal pattern is called a hexagonal number. For example, 1, 6, 15, 28, 45, . . . are hexagonal numbers.

Icosahedron — An icosahedron is a polyhedron with twenty triangular faces.

Infinite (sequence) — A sequence is infinite if it has no last term.

Inscribed Polygon — A polygon is inscribed in a circle if all vertices of the polygon lie on the circle.

Line Segment — A line segment may be described as part of a line. More precisely, segment AB is the set of points consisting of A and B, which are the end points of the segment, and all the points between A and B.

Network — A network is a figure consisting of points joined by lines.

Numerical Coefficient — The numerical coefficient of an algebraic term is the numerical factor of the term. For example, seven is the numerical coefficient in $7x^3$.

Oblong Number — The number of objects which can be arranged in a rectangular array having one dimension one unit longer than the other is called an oblong number.

Octahedron — An octahedron is a polyhedron with eight triangular faces.

Pascal's Triangle — Pascal's Triangle is an infinite triangular array displaying combinations of n things taken n at a time. For example,

$$
\begin{array}{ccccccccc}
 & & & & 1 & & & & \\
 & & & 1 & & 1 & & & \\
 & & 1 & & 2 & & 1 & & \\
 & 1 & & 3 & & 3 & & 1 & \\
1 & & 4 & & 6 & & 4 & & 1 \\
\end{array}
$$

Pentagonal Number — The number of objects that can be arranged in a pentagonal pattern is called a pentagonal number. For example, 1, 5, 12, 22, 35, are pentagonal numbers.

Pick's Theorem — Pick's Theorem states that for geometric figures on a lattice of points, the area equals ½ the number of boundary points plus the number of interior points minus one. Written as a formula it is, $A = \frac{1}{2}N_b + N_i - 1$.

Polygonal Numbers — Polygonal numbers are numbers which can be represented in a polygonal array. For example, triangular, square, pentagonal, hexagonal numbers, etc. are polygonal numbers.

GLOSSARY

Polyhedral Number — The number of objects that can be arranged in a polyhedral pattern is called a polyhedral number.

Polyhedron (plural polyhedra) — A polyhedron is a solid (three-dimensional figure) whose faces are polygons.

Polynomial — A polynomial is an algebraic expression consisting of one or more terms where the terms are numbers or variables or products of numbers and variables.

Prism — A prism is a polyhedron with two faces which are congruent and parallel. All other faces are parallelograms.

Pyramid — A regular pyramid is a polyhedron that has a regular polygon (such as a triangle or square, etc.) for its base and congruent triangles for its other faces.

Pyramidal Number — The number of objects that can be arranged in a pyramidal pattern is called a pyramidal number.

Pythagorean Number Triple — A pythagorean number triple is a set consisting of three numbers a, b, and c such that $a^2+b^2=c^2$. A pythagorean triple is *primitive* if a, b, and c contain no common factors.

Pythagorean Theorem — Pythagorean Theorem states that in a right triangle, the square of the hypotenuse is equal to the sum of the squares of the other two sides.

Sequence — A sequence is an ordered set of numbers or terms formed according to some pattern.

Series — A series is an indicated sum of the terms of a sequence.

Square Number — The number of objects that can be arranged in a square pattern is called a square number. For example, 1, 4, 9, 16, 25, . . . are square numbers.

Square Pyramid — A pyramid with a square base is called a square pyramid.

Triangular Number — The number of objects that can be arranged in an equilateral triangular pattern is called a triangular number. Each triangular number is the sum of consecutive integers starting with one. For example, 1, 3, 6, 10, 15, . . . are triangular numbers.

Triangular Pyramid — A pyramid with a triangular base is called a triangular pyramid.

FIGURATE NUMBERS

Figurate numbers are numbers which can be represented by objects in the pattern of a specific geometric shape. Some selected examples are illustrated below.

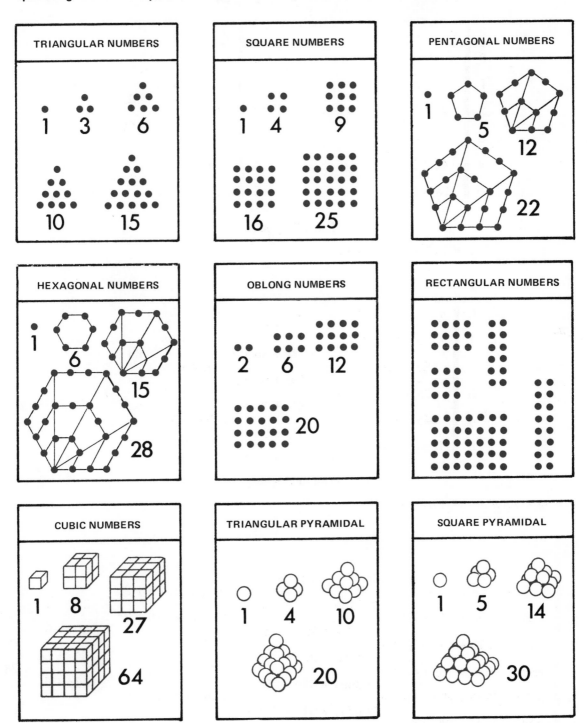

REFERENCE CHART

First Degree

$$y = ax + b$$

x	y		
0	b		
		$> a$	
1	$a + b$		
		$> a$	
2	$2a + b$		
		$> a$	
3	$3a + b$		
		$> a$	
4	$4a + b$		
		$> a$	
5	$5a + b$		

Second Degree

$$y = ax^2 + bx + c$$

x	y		
0	c		
		$> a + b$	
1	$a + b + c$		$> 2a$
		$> 3a + b$	
2	$4a + 2b + c$		$> 2a$
		$> 5a + b$	
3	$9a + 3b + c$		$> 2a$
		$> 7a + b$	
4	$16a + 4b + c$		$> 2a$
		$> 9a + b$	
5	$25a + 5b + c$		

Third Degree

$$y = ax^3 + bx^2 + cx + d$$

x	y			
0	d			
		$> a + b + c$		
1	$a + b + c + d$		$> 6a + 2b$	
		$> 7a + 3b + c$		$> 6a$
2	$8a + 4b + 2c + d$		$> 12a + 2b$	
		$> 19a + 5b + c$		$> 6a$
3	$27a + 9b + 3c + d$		$> 18a + 2b$	
		$> 37a + 7b + c$		$> 6a$
4	$64a + 16b + 4c + d$		$> 24a + 2b$	
		$> 61a + 9b + c$		
5	$125a + 25b + 5c + d$			

Fourth Degree

$$y = ax^4 + bx^3 + cx^2 + dx + e$$

x	y				
0	e				
		$> a + b + c + d$			
1	$a + b + c + d + e$		$> 14a + 6b + 2c$		
		$> 15a + 7b + 3c + d$		$> 36a + 6b$	
2	$16a + 8b + 4c + 2d + e$		$> 50a + 12b + 2c$		$> 24a$
		$> 65a + 19b + 5c + d$		$> 60a + 6b$	
3	$81a + 27b + 9c + 3d + e$		$> 110a + 18b + 2c$		$> 24a$
		$> 175a + 37b + 7c + d$		$> 84a + 6b$	
4	$256a + 64b + 16c + 4d + e$		$> 194a + 24b + 2c$		$> 24a$
		$> 369a + 61b + 9c + d$		$> 108a + 6b$	
5	$625a + 125b + 25c + 5d + e$		$> 302a + 30b + 2c$		
		$> 671a + 91b + 11c + d$			
6	$1296a + 216b + 36c + 6d + e$				

MATHEMATICAL INDUCTION

In studying Finite Differences, our approach to solving a problem is threefold. First we analyze it, then search for patterns, and finally, draw a conclusion or make a generalization based on observation of a limited number of cases, such as the terms of a sequence. This is inductive reasoning, as opposed to deductive reasoning used, for example, to prove theorems in geometry.

Generalizations, such as we have obtained, can be proved to be true for an infinite number of cases by a method called mathematical induction. You will encounter this method later in your study of mathematics. Below is an example of a proof by mathematical induction. This type of proof is extremely useful, but only if the generalization can be found. For certain kinds of sequences and series, the method of finite differences provides this generalization.

Prove that $1 + 4 + 9 + 16 + \ldots + n^2 = \dfrac{n(n+1)(2n+1)}{6}$ is true.

(1) Verify the statement for $n = 1$: $\quad 1 = \dfrac{1 \cdot 2 \cdot 3}{6} = 1 \qquad$ O.K.

Verify the statement for $n = 2$: $\quad 1 + 4 = \dfrac{2 \cdot 3 \cdot 5}{6} = 5 \qquad$ O.K.

(2) Assume the statement is true for any positive integer, $n = k$:
$$1 + 4 + 9 + 16 + \ldots + k^2 = \frac{k(k+1)(2k+1)}{6}.$$

(3) Show the statement is true for the next integer, $n = k + 1$:
$$1 + 4 + 9 + 16 + \ldots + k^2 + (k+1)^2 \stackrel{?}{=} \frac{(k+1)(k+2)(2k+3)}{6}.$$

(4) Add $(k+1)^2$ to both sides of (2):
$$1 + 4 + 9 + 16 + \ldots + k^2 + (k+1)^2 = \frac{k(k+1)(2k+1)}{6} + (k+1)(k+1)$$
$$= \frac{(k+1)(2k^2+k)}{6} + \frac{(k+1)(6k+6)}{6}$$
$$= \frac{(k+1)(2k^2+7k+6)}{6}$$
$$= \frac{(k+1)(k+2)(2k+3)}{6} \text{ , step (3) is true.}$$

It has been proved that if the statement is true for $n = k$, it is also true for $n = k + 1$. It has been shown to be true for $n = 1$ and $n = 2$. Therefore, it is true for $n = 3, 4, 5, \ldots$, in fact, for <u>all</u> positive integers.

BIBLIOGRAPHY

Baker, Betty L., "The Method of Differences in Determination of Formulas," *School Science & Mathematics,* April 1967, pp. 309-315.

Beiler, Albert H., *Recreations in the Theory of Numbers.* 2nd edition, Dover, New York, 1966.

Boole, George, *A Treatise on the Calculus of Finite Differences,* Dover Publications, Inc. New York, 1960.

Brown, Lynn H., "Discovery of Formulas Through Patterns," *Mathematics Teacher,* April 1973, pp. 337-338.

Butts, Thomas, *Problem Solving in Mathematics,* Scott, Foresman and Company, Glenview, Illinois, 1973.

Fitzgerald, Greenes, et al., *Laboratory Manual for Elementary Math,* 2nd edition, Prindle, Weber, and Schmidt, Boston, Mass., 1973.

Gibbs, Richard A., "Euler, Pascal and the Missing Region," *Mathematics Teacher,* January 1973, pp. 27-28.

Greenspan, Donald, "A Finite Difference Proof that $E=Mc^2$," *American Mathematical Monthly,* March 1973, pp. 289-292.

Hemmerly, Howard, "Polyhedral Numbers," *Mathematics Teacher,* April 1973, pp. 356-362.

Rosenbaum, Louise Johnson, *Induction In Mathematics,* Houghton Mifflin Co., Palo Alto, Ca., 1966.

Sawyer, W. W., *The Search for Pattern,* Pelican, Baltimore, Maryland, 1970.

Shklarshy, et al., *USSR Olympiad Problem Book,* W. H. Freeman and Co., San Francisco, 1962, pp. 74-79, 412-22.

This table and the one on the following page are designed to be used with an overhead projector.

QUADRATIC EQUATION $y = ax^2 + bx + c$

x	y	1st Difference	2nd Difference

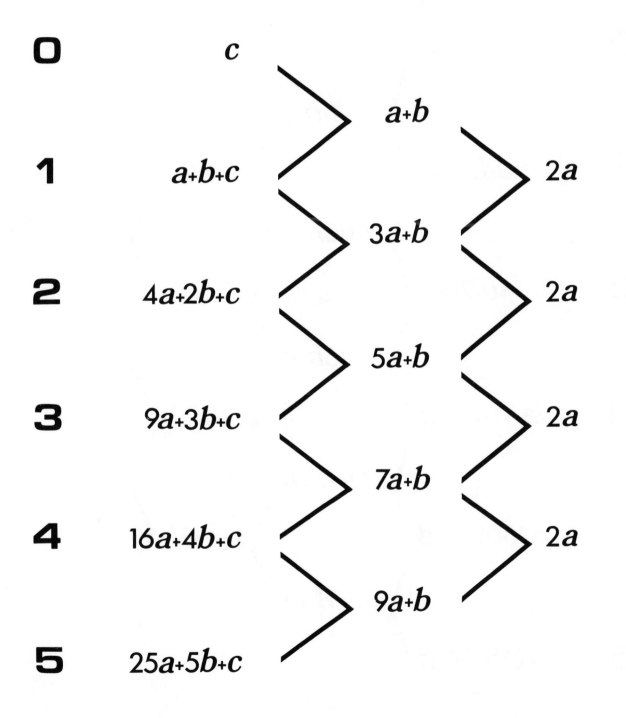

$$0 \qquad c$$
$$a+b$$
$$1 \qquad a+b+c \qquad\qquad 2a$$
$$3a+b$$
$$2 \qquad 4a+2b+c \qquad\qquad 2a$$
$$5a+b$$
$$3 \qquad 9a+3b+c \qquad\qquad 2a$$
$$7a+b$$
$$4 \qquad 16a+4b+c \qquad\qquad 2a$$
$$9a+b$$
$$5 \qquad 25a+5b+c$$

CUBIC EQUATION $y=ax^3+bx^2+cx+d$

x	y	1st Difference	2nd Difference	3rd Difference

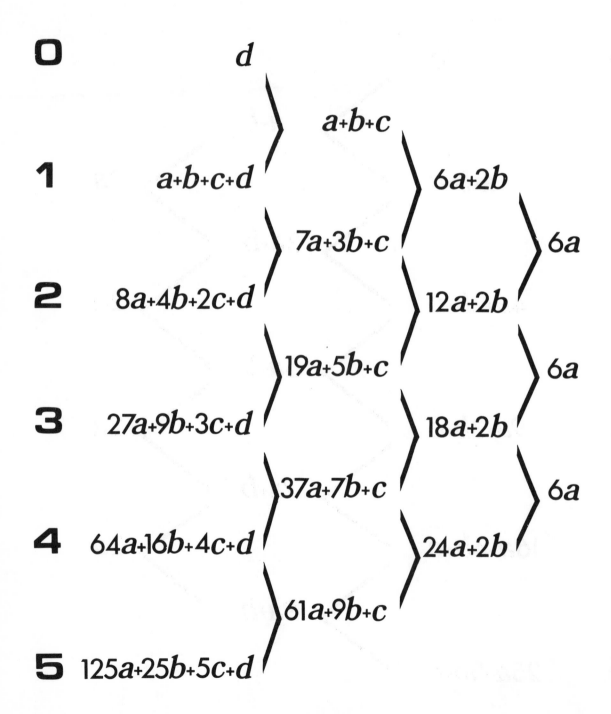

0 d

$a+b+c$

1 $a+b+c+d$

$6a+2b$

$7a+3b+c$

2 $8a+4b+2c+d$

$12a+2b$ $6a$

$19a+5b+c$

3 $27a+9b+3c+d$

$18a+2b$ $6a$

$37a+7b+c$

4 $64a+16b+4c+d$

$24a+2b$ $6a$

$61a+9b+c$

5 $125a+25b+5c+d$